Shipboard Operations

Second edition

H. I. LAVERY

BA, MSc, Master Mariner, MNI, MCIT

Lecturer in Transport Studies
Department of Civil Engineering
and Transport
University of Ulster

Heinemann Newnes

To my parents, Elizabeth and Ernest

Heinemann Newnes
An imprint of Heinemann Professional Publishing Ltd
Halley Court, Jordan Hill, Oxford OX2 8EJ

OXFORD   LONDON   MELBOURNE   AUCKLAND   SINGAPORE
IBADAN   NAIROBI   GABORONE   KINGSTON

First published 1984
Second edition 1990

British Library Cataloguing in Publication Data
Lavery, H. I. (H. Irvine)
   Shipboard operations. – 2nd ed.
   1. Freight transport. Shipping. Ships. Bulk cargoes.
   Cargo handling. Safety aspects
   I. Title
   363.1'23

ISBN 0 434 91091 0

Printed in Great Britain by Redwood Burn,
Trowbridge, Wiltshire.

hipboard Operations

Contents

Preface

In the preface to the first edition in 1984 I wrote, 'I think it would be reasonable to say that the last five years have seen more legislation promulgated with relation to the shipping industry than any other similar period in shipping history'. At the beginning of the following decade I can repeat this statement, as the flow of legislation, necessary though it may be, has continued unabated.

On the International scene the new SOLAS Chapter III is a major item of legislation and implementation of MARPOL 73/78 Annex II and Annex V means more operational changes for ship and shore management. In Britain the 1988 Occupational Health and Safety Regulations and the Ro/Ro Ferry Regulations are major 'packages' which also put increased operational pressures on ships' officers. I have, therefore, completely rewritten Chapter 2 of this book to include all this legislation.

Major changes have also occurred in the educational and examination procedures for Deck Officers studying in the United Kingdom. The BTEC HND in Nautical Science brings nautical education into the mainstream of British education and it is intended that this book will help deck officers of all ranks to achieve that level of proficiency. However, new emphasis has been placed on the responsibilities of Masters and shore management and on the interface between such areas of responsibility. The Chartered Institute of Transport has designated this book a 'fundamental text' for the Maritime Transport paper in the qualifying examinations for membership of the Institute and I have taken cognizance of this.

The main purpose of the book, however, is not to help students pass examinations but to assist management in coping with the bewildering amount of shipping legislation presently in force and in operating ships professionally. There are over 600 IMO and 240 British 'items' of legislation which have some effect on maritime shipping operations: it has been difficult to decide which to include in this book, but I have tried to cover all those that affect the day-to-day operations of a ship.

Most problems at sea are caused by humans rather than by technology. The emphasis in the 1990s must be on improving actual operational practices and it is essential that manning levels should be commensurate with legislative requirements. It is my sincere wish that this book will help management, ashore and afloat, to operate ships in a safe manner. Recently a student from Pakistan told me that, when serving on a Liberian ship, a Norwegian Chief Officer gave him a copy of my book so that he could prepare for a survey: I am pleased to be of such assistance to ships' managers.

Once again, I must thank my wife Sandra and children Sheena and Richard for their forbearance during the many hours that I spent immersed in regulations. Now we can go on the many long walks that I promised you!

H.I.L.

Acknowledgements
Blohm & Voss A G, Hamburg (Figure 9.7).
BP Shipping Limited, London (Figures 5.2 and 5.9).
Butterworth Systems (U.K.) Limited, London (Figures 6.3 aznd 6.4).
Harland and Wolff Limited, Belfast (Figure 9.1).
Jotun-Henry Clark Limited, Marine Coatings, London (Tables 3.2 and 3.3).
Other figures drawn by Sandra Lavery.

Extracts from British Standards are reproduced by permission of the British Standards Institution, Linford Wood, Milton Keynes, MK14 6LE, from whom complete copies of the standards can be obtained.

1

Safety: Operational

Role of the Safety Officer

Since 1 October 1982 the employer of the crew on a United Kingdom ship has been required under The Merchant Shipping (Safety Officials and Reporting of Accidents and Dangerous Occurrences) Regulations, 1982 to appoint a Safety Officer (see Chapter 2).

The duties of a Safety Officer are to:

1 Endeavour to ensure that the provisions of the Code of Safe Working Practices are complied with.
2 Endeavour to ensure that the employer's occupational health and safety policies are complied with.
3 Investigate
    (a) every accident required to be notified by the Merchant Shipping Act
    (b) every dangerous occurrence
    (c) all potential hazards to occupational health and safety
    and to make recommendations to the master to prevent the recurrence of an accident or to remove the hazard.
4 Investigate all non-frivolous complaints by crew members concerning occupational health and safety.
5 Carry out occupational health and safety inspections of each accessible part of the ship at least once every three months.
6 Make representations and, where appropriate, recommendations to the master (and through him to the company) about any deficiency in the ship with regard to
    (a) any legislative requirement relating to occupational health and safety
    (b) any relevant M notice
    (c) any provision of the Code of Safe Working Practices
7 Ensure so far as possible that safety instructions, rules, and guidance are complied with.
8 Maintain a record book describing all the circumstances and details of all

accidents and dangerous occurrences, and of all other procedures required by his duties, and to make the records available for inspection by appropriate personnel.

9 Stop any work which he reasonably believes may cause a serious accident and inform the appropriate officer.

10 Carry out the requirements of the safety committee.

Many mariners consider that the above duties, which are additional to 'normal' duties, place an onerous burden on the officer concerned. However, safety has always been of paramount importance on board ship and some of the above requirements only put into legislation the common practice of efficient seamen.

The Safety Officer needs to be well conversant with the legislation described in Chapter 2 and in particular the Code of Safe Practice for Merchant Seamen, known to seamen as 'The Code'. The Department of Transport has published *Guidance Notes for Safety Officials*; these notes should be carefully studied, particularly those sections which discuss the ramifications of the officer's statutory duties.

The Safety Officer should always be on the lookout for potential hazards and must try to develop a high level of safety consciousness among the crew. This will probably be the most difficult aspect of his job as there can be a high level of safety apathy, and not consciousness, among the officers and seamen. He should aim to become the ship's adviser on occupational safety, which means that the Safety Officer himself must set a high personal standard of safety awareness.

When carrying out the occupational health and safety inspections the Safety Officer must pay attention to the environmental factors as well as to the 'statutory factors'. The galley is a good area to consider. It is very important that the extinguishers are well maintained but it is also important that the air extractor hoods are regularly cleaned to reduce the fire risk from accumulated grease, and that dirt does not accumulate in areas which would produce a health hazard. Thus, the Safety Officer does not only conduct an inspection for the 'safety equipment checklist' but must carry out an environmental inspection to ensure that occupational safety standards are being maintained. Appendix 9 of the guidance notes lists some factors which must be considered, a few of which are noted below.

1 Are means of access to the area under inspection in a safe condition, well lit, and unobstructed?

2 Are fixtures and fittings over which seamen might trip or which project, particularly overhead, thereby causing potential hazards, suitably painted or marked?

3 Are all guard-rails in place, secure, and in good condition?

4 Are lighting levels adequate?

2

5 Is ventilation adequate?
6 Is machinery adequately guarded where necessary?
7 Are permits to work used when necessary?
8 Is the level of supervision adequate, particularly for unexperienced crew?

The investigation of accidents and dangerous occurrences will be an important part of the Safety Officer's duties. The actual reporting of an accident will be carried out by the master but it is the statutory duty of the Safety Officer to investigate the incident and to assist the master to complete the accident report form. The first form was issued in September 1982, *Form ARF/1*, and the explanatory notes which accompany the form should be closely followed.

The Safety Officer should have a chat with the ship's personnel to explain the purpose and function of the form and to dispel any misconceptions to which the 'galley radio' may have given rise. The following points should be emphasized:

1 The purpose of the form is to ascertain the causes of accidents and ultimately to reduce the chances of a similar accident happening again.
2 All forms are treated in the strictest confidence by the Department of Transport.
3 None of the forms will be used by the Department of Transport in a prosecution or an investigation.
4 The form is computer processed and thus will be seen by very few people.
5 Personal names are deliberately omitted from the form and thus anonymity is ensured.

Officers should be aware that the post of Safety Officer is not a sinecure and that much effort should be put into the role in order to meet the obligations required by the regulations.

Role of the safety representative

In every ship to which the regulations apply the officers and ratings may elect safety representatives, but are under no obligation to do so. However, it would be remiss not to do so and:

> in ships carrying fewer than 16 crew, one safety representative may be elected by the officers and ratings; in ships carrying more than 15 crew one safety representative may be elected by the officers and one safety representative may be elected by the ratings.

The safety representative has powers but no duties, and he may:

1 Participate in any of the inspections or investigations conducted by the Safety Officer, provided that the latter agrees to such participation.

3

2 Undertake similar inspections or investigations himself, providing that notification of such activities has been given to the master.
3 On behalf of the crew on matters affecting occupational health and safety
   (a) consult with the master and the Safety Officer and make recommendations to them, including recommendations to the master, 'that any work which the safety representative believes may cause an accident should be suspended';
   (b) make representations through the master to the employer;
   (c) request through the safety committee an investigation by the Safety Officer of any such matter.
4 Inspect any of the Safety Officer's records.

The safety representatives need to develop a good relationship with the Safety Officer and should work with him to raise safety standards. The spirit, and the purpose, of the regulations would be badly damaged by representatives who might use their powers as a 'negotiating weapon' in any dispute with employers. The role of the safety representative should not be abused by personnel who wish to use the post in an obstructive, instead of a constructive, manner. The employer has an obligation to formulate rules for the election of safety representatives and thus elections should take place. The posts should not be filled by persons who are only nominated, either by persons or unions, as the regulations make it clear that the posts can only be filled by elected personnel.

Safety committee

If safety representatives are elected on any ship the employer must appoint a safety committee, i.e. safety committees are mandatory on any ship which has elected safety representatives. However, it would be a wise practice to institute a safety committee on all ships. The membership of the committee must include the master as chairman, the Safety Officer, and every safety representative. The duties are to:

1 Ensure that the provisions of the Code of Safe Working Practices are complied with.
2 Improve the standard of safety consciousness among the crew.
3 Make representations and recommendations on behalf of the crew to the employer.
4 Inspect any of the Safety Officer's records.
5 Ensure the observance of the employer's occupational health and safety policies.
6 Consider and take any appropriate action in respect of any occupational health and safety matters affecting the crew.

7 Keep a record of all proceedings.

A well-organized committee, which meets regularly, can be of great assistance to those entrusted with safety. Minutes should be kept, with copies posted on the ship's notice boards and a further copy sent to the company's office. In addition to the safety representatives, personnel from all 'sections' of the ship should attend, e.g. cadets, petty officers, stewards, etc. The committee should not be dominated by senior officers and efforts should be made to encourage the junior ranks attending to put forward their ideas. Reports should be presented stating the maintenance and drills that have been carried out since the previous meeting. Once a project or idea has been accepted by the committee it must be put into action, otherwise the committee loses impetus and members will regard it merely as a sop to company and Merchant Shipping regulations and of little practical use. Members should study an advisory booklet published by the General Council of British Shipping, *Accident Prevention Organisation on Board Ship*, and the advice to safety committees in the Department of Transport guidance notes.

The committee should be the safety forum on board ship and safety should be its only concern. It should not become involved in discussion on 'conditions of service' or trade union matters.

Methods for improving and maintaining the safety awareness of crews

Maintaining the interest of a crew in all aspects of safety can be a difficult and, at times, frustrating and unrewarding task. This list contains suggestions which the Safety Officer could employ in order to promote safety awareness. I know from experience that many sailors are extremely lax in adhering to safety requirements. However, some of the following methods have been used on board ships to good effect. They should be regarded as practical ideas and not just 'waffle' to be regurgitated in order to pass an examination.

Films

An extremely useful method on those vessels which carry projectors and other viewing equipment. Various organizations, commercial or otherwise, produce safety films which can be borrowed or hired. Experience indicates that the best time for showing educational films is immediately preceding feature films on those ships fortunate enough to have such a service.

Posters

This can be an effective method of bringing particular dangers to the attention of crew members. Posters should be situated in those spaces where the danger warnings are most pertinent and should be changed frequently before they become part of the furniture and thus ignored. The placing of posters within living or recreational areas is a contentious issue; many seamen believe that it detracts from the 'quality of living' in that area of the accommodation. Posters can be obtained from the General Council of British Shipping and other sources.

Publications

A number of useful booklets have been published by the Department of Transport and copies should be given to the crew. These include *Personal Safety on Ships, Personal Survival at Sea,* and *Fire in Ships.*

A small booklet entitled *Safe or Sorry?* is published by the Marine Society; it is worth reading. The General Council of British Shipping issues a good magazine, *Your Safety Aboard Ship,* and copies should be distributed throughout the ship.

Informal talks

Talks in the crew's mess have been found to be a useful method for explaining sections of the 'Code'. The above booklets could be used as the basis of such chats. It may be useful to talk to sections of the crew, e.g. the catering staff, who often have less safety awareness than other crew members.

Maintenance of safety equipment

Involve as many people as possible in the maintenance of safety equipment. This practice emphasizes the fact that safety is the responsibility of everyone on board. There is no reason why stewards, for example, should not be instructed in methods of refilling the extinguishers in the catering area.

Audio-visual aids

Several commercial firms active in producing training aids have good cassettes pertaining to safety. These can be used as an introduction to informal talks, as an aid to maintenance, to assist in the training of emergency teams, or simply as television films.

Fire patrols

Read M notice 528. Patrols, or equivalent inspections, must be carried out at all times whether at sea or in port. Particular attention should be paid to patrolling the accommodation between 2300 hours and 0600 hours. Safety awareness is increased if the patrol is instructed to observe any safety infringements, such as loose chairs, and not to be concerned solely with fire prevention.

Marine safety cards

Published by the General Council of British Shipping, these cards highlight particular dangers on board ship. Card 1 deals with entry into enclosed spaces.

Accident records

Details of accidents should be posted on notice boards as an accident prevention aid. The name of the unfortunate person involved should be withheld.

Days without accident board

It is a common practice for factories, oil terminals, etc., to post notices stating the number of days since the occurrence of the last accident. It might also be useful to do this on board ship.

Safety quiz

This could be open to individuals with a suitable prize being awarded, or to teams representing the several departments on board. This type of quiz has been popular on several ships and the quiz in *Your Safety Aboard Ship* might be used for such a purpose.

'Permit to work' system

This will be discussed in a separate section but it must be explained to the crew and the importance of strict compliance with the permit should be emphasized.

Aspects of the maintenance of safety equipment

The maintenance of safety equipment must be given a high priority. A highly trained efficient emergency team can fail to carry out a task if a key item of equipment is inoperable. It is a basic principle of safety that all equipment must be maintained in excellent condition and be kept available for immediate use at all times. It must be pointed out that it is an offence under the Merchant Shipping Acts for life-saving appliances to be in a defective condition and that a ship may be detained until the defects have been rectified.

Some companies supply books which contain a full list of all emergency equipment on board. If this does not apply to your ship, then a comprehensive list of such items must be made in order that no piece of equipment is overlooked. The safety maintenance plan should be integrated with the ship's 'Planned Maintenance Schedule', but in addition tests should be carried out during routine emergency drills. The areas where the drills take place should be rotated on a carefully planned basis so that all equipment is used at frequent intervals.

The following list does not contain all the safety equipment which the vessel is required to carry. The main aspect is to bring to the attention of the Safety Officer some legal requirements or suggestions of which he may not be aware. It does not contain a full list of safety M notices.

Lifeboats

The minutiae of lifeboat maintenance will not be considered. However, glass-fibre boats should be checked monthly for softness. From 1 May 1981 all morphine should have been removed from survival craft and non-addictive pain-killers substituted for the morphine. M1248 'Automatic Release Hooks for Liferafts and Disengaging Gear for Lifeboats and Rescue Boats' should be studied.

Lifeboat falls

Turned end-for-end at intervals of not more than 30 months and renewed at intervals of not more than 5 years. They should be greased at frequent intervals and regularly inspected for broken strands. Sailors sometimes ignore the sections of falls within blocks. Slack the falls and grease those portions of the wires. Lead blocks should be greased every two weeks and overhauled every six months. Check that the fall becket is secured to the drum end.

Lifeboat davits

Test the limit switch at boat drills. Trackways should be scaled and properly coated with grease as necessary, and pivot points should be greased every fortnight. The main body of the davit should be checked for rust, harbour pins greased, and the wire span for the boarding ropes checked. Inspect the gripes for rust and broken strands. Some gripes are plastic coated; this is rather a dubious practice as once the wire starts to rust the process cannot be stopped. Test senhouse slips and bottlescrews. The brake mechanism should be tested at four-monthly intervals; lower the boat to the water, raise it a metre, apply the brake and see if it holds. Test the hand gear safety device at each drill. Check all ropes for rotting, check that the boarding ladders are secured to the eye pads, and apply pilot ladder maintenance principles to the boarding ladders. M1186 deals with lifeboat winches fitted with a roller ratchet mechanism; winches for lifeboats which are heavily used should be opened and thoroughly examined every 2 years, otherwise to be examined every 4 years.

Liferafts

Merchant shipping regulations require inflatable liferafts to be surveyed at intervals not exceeding twelve months. This can only be carried out at Department of Transport approved service stations (refer to M notices for such stations). Square rigid liferafts may be serviced on board ship by the manufacturer. M1047, 1173 and 1211 should be read in full as they contain much pertinent information regarding inflatable liferafts. Since 25 May 1980 raft lashings must be fitted with an approved automatic release system of a hydrostatic or equivalent nature, and rafts must be stowed in such a manner that they float free from the vessel in the event of sinking.

Lifebuoys

Check for cracks and, if any are found, replace the lifebuoys and have the old ones destroyed ashore. When necessary repaint the name and port of registry. Inspect the grablines and connections. Two of the buoys must have a 27.5 m buoyant line attached; ensure that these lines have not become too worn or tangled. Half of the lifebuoys must be provided with self-igniting or self-actuating lights; inspect the lights regularly. Frequent inspections should be made of the two quick release buoys on the bridge, especially the lines between the buoys and the light and smoke signals. Make sure that any releasing pins, swivels, etc., are well greased and are not frozen. Lifebuoys must weigh at least 4.3 kg if their weight is used to release light or smoke signals.

Lifejackets

Donning instructions should be displayed in conspicuous positions. The report of the loss of the *m.v. Lovat* recommends that 'an additional supply of lifejackets should be kept in some such position as the bridge to cater for situations in which it is not possible for all the crew to collect their lifejackets from their accommodation'. M1238 gives the recommended scale of the number of additional lifejackets, e.g. if the vessel is certified to carry more than 16 persons additional lifejackets for not less than 25 percent of the certified number are required. The jackets should be stowed near the normal embarkation locations in a suitable dry, unlocked and marked position.

Retro-reflective material

This should be fitted on lifeboats, liferafts, lifebuoys, buoyant apparatus and lifejackets. M1056 gives the details.

Cordage

Safety Officers should be aware that a cordage table for life-saving appliances is contained in M1232. This should be referred to when renewing lifeboat grablines, etc.

International shore connection

Ensure that the securing bolts are free and keep them well greased. If possible, a connection should be stowed in such a position that it is not exposed to the elements. Although only one connection is required by the regulations, some Chief Officers on VLCCs (very large crude carriers) find it a good practice to have three located about the ship, one under the focsle, one near the gangway location, and one aft. A spare one could be kept at the Emergency Squad mustering station. Specifications for a connection can be found in the regulations and if friendly relations have been maintained with the Second Engineer a few can be made on board. The location of the connections should be clearly indicated.

Emergency fire pump

To be tested weekly. Good standards of seamanship should be the prime factor in maintaining equipment. However, a defective emergency fire pump can involve the detention of a ship until it is repaired.

Fire hoses

Canvas hoses should always be dried after use before being stowed. Failure to do so will result in rot which first shows up by pinhole leaks in numerous places along the hose. If this occurs the hose must be replaced. Hoses other than canvas can also rot and crack and must be inspected frequently. The hoses should be frequently tested in the rotational system of emergency drills.

Male and female couplings can be damaged easily by dropping. If they do not connect properly with either the hydrant or nozzle, they must be renewed immediately.

Hose boxes should be maintained in a clean, well-painted condition. Hinges, etc. should be checked for rust and kept greased. The hose number should be clearly indicated on the boxes. Do not stow other equipment or rags in the boxes and remember to check the hoses within the accommodation.

Fire hydrant valves

These should be kept well greased and when possible should be checked every week to see that they are free and do not require a wheel spanner to make them turn. Check the accommodation hydrants; on a newly built ship it was found that several valves had been incorrectly fitted and they would have been useless if needed to fight a fire in that area.

Fire nozzles

These are liable to misappropriation and theft, especially in port. Although it is difficult, the Safety Officer must try to ensure their security. Inspect periodically under pressure to ensure that the nozzles operate satisfactorily. Any scratches or indentations on the inner surface will spoil a jet. Inspect also for general damage, putting emphasis on the mechanism of dual-purpose nozzles.

Fire extinguishers

Portable extinguishers are the first line of defence. Most fires have small beginnings and prompt action with an extinguisher can often deal with an emergency. All extinguishers should be:
(a) located in an easily accessible position
(b) the correct type to deal with the class of fire expected in that area
(c) painted in the appropriate colour code
(d) regularly inspected and tested
  The colour coding should be:

Water	Signal red
Foam	Pale cream
Powder	French blue
Carbon dioxide	Black
Halon	Emerald green

The European Standard fire classifications are:

A A fire involving solid materials usually of an organic nature, e.g. wood, cloth, paper.
B A fire involving liquids or liquefiable solids, e.g. hydro-carbons such as petrol, kerosenes, cooking oils.
C A fire involving gases, e.g. from coal, fermenting sugar.
D A fire involving combustible metals, e.g. sodium, potassium, magnesium.

  The best media for extinguishing the classes are:

A Water, dry powder.
B Foam, dry powder.

C Carbon dioxide, halon.
D Presents a problem to mariners, especially if the superstructure is made of aluminium. If water is used against fires involving metals, a violent reaction can take place which may result in the spreading of the fire and/or explosion. Powder extinguishers appear to be most effective but it would seem that no one method can deal with all metal fires.

The date of testing and refill should be clearly marked on the extinguisher. Dry powder extinguishers should be shaken to ensure that the powder is moving freely as the powder can 'cake' inside the cylinder, especially in humid areas. Check that extinguisher nozzles are not blocked. M765 states that non-portable extinguishers which operate by being rotated 90 degrees from the vertical have been secured in such a manner as to require spanners to make the operation possible. It is common sense to ensure that all extinguishers are immediately available.

Breathing apparatus, smoke helmet or mask type

Inspect the seal on the face mask for signs of perishing on the rubber and ensure that the visor is undamaged. Inspect the air hose for damage, especially around the couplings. Wash and dry before restowing. Clean and oil the air pump or bellows and check the protection on the air inlet. Test the bellows or pump before use.

Breathing apparatus, self-contained

As above, check the mask seal and clean the face piece. Inspect any threaded fittings for damaged threads and obstructions. Check main and by-pass valves and inspect the pressure gauge for visible damage. Check the pressure of the operating and spare air cylinders; any serious fall in pressure can be rectified if the vessel has charging facilities (each ship should have a portable compressor so that cylinders can be used frequently in training exercises). Otherwise send them ashore for recharging; spares should be supplied while that operation is being carried out.

Breathing apparatus, both types

Read the manufacturers' manuals and ensure that operating instructions are attached to the apparatus. Each apparatus should have:
(a) a fire-proof life-and-signalling line
(b) an adjustable harness
(c) means for protecting eyes and face
(d) signal plates on the harness and the free end of the lifeline
The complete unit, including the harness should be occasionally wiped down with a mild disinfectant solution. The wiping of the inside of the face

mask with a solution of dish-washing liquid often prevents it fogging up when in use. Inspect the harness and line for signs of wear and damage and clean the apparatus case.

Fireman's outfit

This should contain:
(a) a breathing apparatus
(b) a portable self-contained electric battery operated safety lamp of three hours' duration
(c) a fireman's axe
(d) protective clothing
(e) boots and gloves
(f) a rigid helmet

The outfits should be stowed in accessible positions not likely to be cut off by fire. It is recommended that one should be located at the Emergency Team Assembly Station.

Fixed gas fire extinguishing installations

Ensure that the outlets to the protected areas are open and painted red to identify them as fire-fighting appliances. Inspect all control valves and cocks and check the permanent marking which indicates the compartments to which the pipes are led. Personnel should be aware of the nature of the audible warning and such an alarm should be checked regularly. The access doors to carbon dioxide storage rooms should be gas-tight and insulated and the rooms should be clearly marked in permanent lettering. The contents of the cylinders should be ascertained by weighing or by the isotope method. The cylinders should also be visually inspected and any showing signs of pitting should be replaced as soon as possible. M681 warns that cylinders in gang release systems have rotated and made the system inoperable due to misalignment of the valve operating levers; the alignment of the cylinders should be checked regularly. M825 reports that loss of life has occurred when carbon dioxide systems have been activated accidentally during repair periods or in normal service. Sufficient measures were not taken to guard against accidental release or to issue suitable instructions regarding the operation.

Additional safety equipment

M1027 states that on some ships safety equipment which is excess to statutory requirements is not being maintained properly. All equipment must be maintained to a high standard so that it can be used in any emergency which might arise.

Emergency station lists

M1217 which deals with musters and drills should be studied in conjunction with the Merchant Shipping Musters and Training Regulations 1986. The information which is required to be shown on muster or emergency station lists is detailed in the above documents, as is the frequency of drills and musters. It should be emphasized, however, that teams should train as frequently as possible and not just carry out the minimum requirements.

The trend in emergency training over recent years has been to be as realistic as possible and this has been reflected by the use of the 'Emergency Team' system. This system divides the crew into teams which should be trained in their various functions so that when an emergency occurs the whole crew swings into action. This is not an idealistic picture; it is possible. The crews of tankers and other specialized vessels have been using the system for years and have shown that a willing crew, led by enthusiastic officers, can be trained to a good standard. Thus, when the emergency alarm sounds one should observe seamen moving quickly with a purposeful air to their allotted stations and not, as in some cases, huddling in small groups unsure of what they should be doing.

The number of teams into which the crew should be divided and the exact membership of such teams is often a matter for an individual Safety Officer's preference and experience.

If the overall system works one should not lay down strict guidelines for the composition of teams. Even the names of the teams will vary from ship to ship. A suggested system is:

Overall Control Team (on bridge)	*Support Control Team*
Master	*(at engine room controls)*
Radio Officer	Chief Engineer
Rating	Fourth Engineer
Deck Cadet	Junior Engineer
	Electrician
	Junior Petty Officer

Emergency Team	*Back-Up Team (on poop)*
(Emergency Team H.Q.)	Second Officer
Chief Officer	Third Engineer
Second Engineer	Junior Engineer
Chief Petty Officer	Petty Officer
Senior Rating (Deck)	Senior Rating
Senior Rating (Engine Room)	Rating (Deck)
Rating	Rating (Engine Room)
Rating	Cadet
Cadet	

Reserve Team (boat deck)
   Third Officer
   Catering Officer
   Senior Rating (Deck)
   Chief Cook
   Second Cook
   Steward
   Steward

The function of the various teams is given below.

Control team

The Captain with his team co-ordinate all operations, maintain internal and external communications, keep records, and plan the overall strategy as the emergency develops.

Support Control Team

The Chief Engineer is in charge of the engine room and is responsible for maintaining normal and emergency services. He will also advise the Master on mechanical matters and if necessary take charge of engine room emergencies.

Emergency Team

This team deals with the emergency at source. On arriving at the scene of the incident, the team takes the necessary corrective action and informs the Control Team of the situation. The Chief Officer is in charge of 'deck' emergencies, but if the incident occurs in the engine room the Second Engineer leads the team.

Back-Up Team

Most incidents can be dealt with by the Emergency Team. The Back-Up Team may be needed to provide support by carrying additional equipment to the first team, to provide stretcher bearers, and to relieve injured or tired men.

Reserve Team

Upon assembling on the boat deck, if weather conditions permit, this team should clear and prepare to launch the lifeboats and liferafts. If the incident poses a danger to the ship, the boats are then ready to take injured and uninvolved persons such as wives. If the incident is minor, then the Reserve Team will have benefitted from preparing the boats in an emergency.

The Emergency Team should be kept as small as possible; an eight man maximum is suggested. Some vessels have a Team 1 and a Team 2 of equal standing, led respectively by the Chief Officer and the Second Engineer, but on some occasions overcrowding at an incident has occurred. The smaller the number, the easier it is to train to a high standard. Once a good standard has been reached men can be interchanged between teams. On many vessels the manning standards may be such that there are only sufficient men for one emergency team.

One should remember that Lifeboat Muster Lists should also be displayed and Muster Drills held so as to ensure that everyone knows their Abandon Ship procedure. An individual's emergency stations and duties should be put on a card which is displayed in that seaman's cabin.

Emergency Team training

The type of drill should be varied and all the safety equipment should be used at regular intervals. The team members should be able to use all equipment, e.g. the breathing apparatus, but their individual duties at each type of incident must be carefully explained and rigidly adhered to. The basic training of the team should not involve the whole crew. However, the overall system should be explained to all personnel and frequent exercises involving all the crew should be carried out so as to prevent a 'them and us' situation developing. All the crew should be trained in lifeboat and liferaft operations.

The timing of drills should be announced but not the type of incident; thus the maximum degree of reality is obtained.

If possible an Emergency Headquarters should be designated; a room or large locker on the external perimeter of the accommodation with internal and external access is best. Some of the ship's statutory and additional equipment should be stowed there so that some of the gear required for all emergencies can be quickly put into operation. After the emergency signal has been sounded the type of emergency should be announced over the public address system. Thus if the Chief Officer has been delayed by his operational duties, he will find the team fully equipped and ready to go when he arrives at the headquarters.

It is important to have exercise post-mortems and the headquarters can be used for such discussions. Talks and demonstrations can also be given there. Although fire is one of the main hazards to mariners it should be emphasized that not all emergencies involve fire. Officers have arrived at the headquarters to find the team wielding extinguishers and hoses when the drill was 'man overboard'. Incidentally, such a drill can be carried out easily during a period at anchor. Other types of drill can be accommodation, hold and engine room fires, accidents in enclosed spaces, helicopter crash on deck, injured seaman fallen from a height, tank explosions, oil pollution and collision damage. An exercise in which the whole crew can join, and which does have an element of fun, is one in which crew members are blind-folded outside their own cabin and are then instructed to proceed to their emergency station. Although this exercise does produce a few laughs, it never fails to show the difficulty of moving through seemingly familiar surroundings in total darkness or smoke-filled conditions.

Some seamen and shore managers often regard time spent in safety training as lost time, especially on ships with reduced manning levels. Fortunately such members of administrative shore staff are now realizing that a few hours a week spent on training can save many lives and millions of pounds. On board ship a few senior officers still only pay lip service to the ideals of efficient safety prevention and training. Indeed, some of them only train in order to 'keep the company happy'. Senior officers must give enthusiastic support to the Safety Officer's work.

To conclude this section I would strongly suggest that crews should be given a talk on the inadvisability of prematurely abandoning the vessel. In nearly all emergencies the ship is the safest place and case histories of shipping disasters have shown that many people have died abandoning vessels which did not eventually sink. Seamen are excessively concerned about the dangers of undertow or being struck from below by surfacing wreckage. These are minor when compared to the danger from injury during abandonment and from exposure during the subsequent period. Such a talk should include a section on abandonment preparation if that drastic step becomes necessary. In cold climes the effect of quick immersion, known as 'cold shock', may prove to be disabling or even fatal. Before donning a lifejacket several layers of warm woolly clothing and an anorak should be put on. The extra clothing will prolong survival time as it will reduce heat loss and the air trapped in the layers of clothing will aid flotation. All mariners should endeavour to attend the two-day 'Survival at Sea' course which most nautical colleges offer.

M notices which help with training are 1118, 1204, 1206, 1218 and 1267.

Emergency drills

Many drills can be performed and each drill must be slanted to the requirements and layouts of particular ships. The following are two which might be of particular interest.

Abandonment by liferaft

This drill is often neglected. I personally believe that launching a lifeboat in wind conditions over force 5 is a very risky operation and that the best chance for survival lies in the proper use of liferafts. M1217 reminds mariners that on vessels with davit-launched liferafts, one liferaft should, if possible, be inflated and lowered at least once every four months in port. The Department of Transport 'strongly recommends that drills in general should include some preparation for abandonment into liferafts'.

Such drills must take into account the characteristics of the particular ship involved and there will probably be a need for disembarkation points separate from those for lifeboats. The necessity for such drills has been highlighted by a recent abandonment in which a liferaft was inflated on deck. It is contrary to seamens' instincts to criticize the actions of fellow mariners when their lives are in extreme peril but good training will help to prevent such obvious blunders.

Tests by the National Maritime Institute have shown that rafts are at their most vulnerable when just launched with no one aboard. It is important quickly to get a few men into the raft and to have them sit on the windward side so as to make the raft stable enough for others to board. It is also important to stream quickly the sea anchor in order to reduce drift and to aid stability. Davit-launched rafts are particularly difficult to control on high-sided vessels and seamen should be reminded that boarding all types of raft can be a hazardous experience.

Fire fighting in port

Although many fires occur in port it can be difficult to arrange a drill with the local fire authorities. This problem can be partly resolved by instructing the team on shore requirements. All ships should have an updated fire wallet containing the following information:
(a) a general arrangement plan
(b) a ventilation plan
(c) a shell expansion plan in case it will be necessary to cut through the ship's side
(d) a plan of the fire-fighting equipment
(e) electrical data
(f) stability data due to the danger of free surface and other effects
(g) a cargo plan with any dangerous cargoes being specifically mentioned
(h) location of watertight doors and fire-resistant partitions
(i) any drilling machines and special equipment that the vessel carries
  The Senior Fire Officer should be presented with the wallet on his arrival. He will also probably require the following information:
(a) the exact location of the fire and the chances of it spreading to other compartments
(b) contents of double bottoms or deep tanks in the vicinity

18

(c) what the ship's staff are doing and how many hoses and pumps are in operation
(d) if any fixed fire-fighting installation is in operation
(e) the state of cargo operations
(f) the condition of fuel oil, ballast and fresh water tanks
(g) the ship's communication systems
(h) the number of people on board
(i) any peculiarities of the ship's design

M1267, *Fire Prevention and Fire Fighting in Ships in Port*, should be studied. A joint drill with shore authorities should be arranged.

Care and maintenance of ropes

Although this is very much the province of the Chief Officer, knowledge of correct rope usage will help the Safety Officer in his role of hazard spotting and accident prevention.

Natural fibre ropes such as manila, hemp and sisal have been replaced largely by man-made fibres, although mariners still prefer the use of manila for gantlines.

Nylon The strongest of man-made fibres, it has high energy absorption and the ability to endure heavy repeat loading. Unfortunately it sinks.

Polyester Has good abrasion resistance and a lower extension than most synthetic ropes. It has equal strength when either wet or dry but being heavier than nylon it also sinks.

Polypropylene The most common type of mooring rope due to its ability to float. It is of equal strength wet or dry.

Care and handling of synthetic fibre ropes

1 As all synthetic ropes have varying qualities of resistance to chemicals, acids, alkalis, solvents, etc., they should be stowed in well-ventilated dry compartments away from such materials.
2 Do not stow on deck, even for short passages between ports.
3 In port, cover coiled mooring ropes as exposure to strong sunlight is detrimental to ropes.
4 Stow on gratings to avoid inadvertent contamination.
5 Do not stow near heat, e.g. engine room bulkheads.
6 Inspect regularly for internal and external wear and tear. In cases of excessive wear, powdering will be visible between the strands. Remember that synthetic ropes often become 'dosed' internally while looking good externally. Overworked ropes may become hard, stiff, and hairy.

7 Eye splices should have not less than four tucks. The splice should then be tapered by halving and quartering the strands for two tucks respectively. The tapered part of the splice should be securely wrapped with adhesive tape.

8 Synthetic ropes should be of a type providing a grip similar to that of natural ropes.

9 Avoid overloading the rope around sharp angles.

10 Never put strain on a kink as it can cause permanent damage. The visual effects of such damage may be removed but a loss of strength of one-third can be experienced in the kink area.

11 Wash the rope with fresh water in the event of it being splashed by corrosives.

12 Where wire is to be joined to a rope, ensure that a thimble is used and the wire and rope are laid the same way.

13 Keep wires and ropes in different leads.

14 Sections of ropes which are vulnerable to abrasion, e.g. the eyes, should be protected with leather sheaths.

15 Do not cross cut ropes on drums.

16 Synthetic ropes have low melting points, therefore do not surge or render on drum ends. Do not use more than three turns.

17 Always stopper off with the same type of rope using a 'West Country' or 'Norwegian' stopper.

18 Try to prevent mooring ropes from snagging on quays or on cargo ashore.

19 New coils should be unwrapped in an anti-clockwise direction from the coil centre, or the coil should be suspended by a swivel and the rope taken from the outside.

20 Seamen should be warned that there is no audible warning when a synthetic rope is approaching breaking point. Seamen have been decapitated by the whiplash action of such ropes.

Advantages and disadvantages of synthetic rope

Advantages

1 High tensile strength.

2 Good durability as it is less prone to gradual loss of strength.

3 Resistant to rot and mildew.

4 Stretches more than natural fibres.

5 Does not kink easily and if constructed in plaited lay does not readily open up.

6 Smaller than natural fibre ropes for same strength.

7 Easy to handle.

8 Does not become less pliable with age unless overworked.

Disadvantages

1 Due to the ability to stretch, there is a considerable whiplash effect if the rope breaks.
2 No audible warning prior to breaking.
3 Low melting point, therefore it has a tendency to melt or fuse on the drum end.
4 Susceptible to heat and sunlight.
5 Can be contaminated by chemicals, etc., and thus weakened considerably with little visible evidence of such.
6 Plaited ropes require special splicing instructions.

Natural fibre ropes

Manila should be used for pilot ladder construction, some life-saving appliances, lizards, etc. The care is similar to that for synthetic ropes. However, more turns can be put on the drum end.

Advantages

1 Do not melt.
2 Give audible warning if breaking point is approaching.
3 Do not recoil as much as synthetic when broken.
4 Not susceptible to moderate heat and sunlight damage.
5 Can be surged and rendered on drum ends.
6 Can be opened up for internal inspection without damaging the rope.
7 Easily spliced.

Disadvantages

1 Susceptible to rot and mildew.
2 Not as strong as synthetic ropes of the same size.
3 Has small stretching ability.
4 Not easily handled. Has a tendency to swell and stiffen with age and damp which makes large mooring ropes difficult to work with. If wet can freeze in very cold conditions.

Chapter 15 of the 'Code' should be studied in full for the care and inspection of gantlines used with bosun's chairs, safety harnesses, and stage ropes. The safety of seamen using the above appliances depends very much on the conditions of the ropes and they must be given a high degree of care and attention. Particular attention should be paid to the following points:

1 Such ropes should be stowed in a special locker and should be used for no other purpose. Nothing else should be stowed in the locker.
2 All gantlines should be clearly marked for their particular function, e.g. funnel, bridge front.
3 Make sure the splices are correct.
4 All blocks and lizards should be in the same condition as the gantlines.

5 A palm and needle whipping should be on all gantline ends.
6 All gantlines should be thoroughly inspected each time before use and daily when in use.
7 The ropes must be load tested before use to four or five times the weight which they will be required to carry.

Also read all M notices relating to rope safety, e.g. M718 on mooring equipment and M1336 on tows.

Safety precautions in the care and handling of batteries

This subject provides a useful talk for the Safety Officer.

Particular hazards when charging batteries are hydrogen explosion and short circuits. During charging a battery gives off hydrogen and oxygen and the subsequent mixture can be easily ignited. Short circuits may cause arcing which could initiate an explosion or burn the operator.

General precautions for all batteries

1 Compartments should be well ventilated to prevent any build-up of dangerous gases.
2 A 'No Smoking/No Naked Lights' sign should be displayed on the outside of the door to the compartment and also inside the compartment.
3 The compartment should be used for batteries only and not for odd pieces of equipment such as NUC signals.
4 The compartment light bulbs should be protected by gas-tight covers and all wiring should be well insulated.
5 All battery connections should be clean and tight.
6 The batteries should be securely stowed.
7 Metal tools should never be placed on top of batteries as they may cause short circuits.
8 For the same reason, rings should not be worn when working with batteries as short circuits may cause burns.
9 When being moved batteries should be carried horizontally. As they are very heavy the batteries should be carried by sufficient personnel and in such a manner as to avoid injury. Liquid spills can cause corrosive injuries and damage to clothing.
10 All battery circuits should be dead when leads are being connected or disconnected.
11 The battery compartment should be kept locked with an emergency key in a glass box beside the entrance.
12 Do not use portable electrical equipment in the compartment.

Precautions when handling lead acid batteries

1 Sulphuric acid is hygroscopic, i.e. it has a great affinity for water. When preparing the electrolyte the acid should be slowly added to the water; if the water is added to the acid the heat generated could cause an explosion which would spray sulphuric acid over the handler.
2 Protective clothing such as goggles and rubber gloves should always be worn.
3 The terminals should be protected by petroleum jelly. The deposits in the area of the terminals can be injurious to eyes and skin.
4 Do not use an excessive charging rate as an acid mist may come out of the battery vents, settle on to adjacent surfaces, and cause burns to operators.
5 In the event of accidents the acid should be neutralized with copious amounts of water. Eyewash containers and a supply of fresh water should be readily to hand. The container should be distinguishable by touch.

Precautions when handling alkaline batteries

1 The metal cases of these batteries are 'live' and should not be touched by the body or with tools.
2 The electrolyte is corrosive and in the event of accident should be neutralized with boracic powder solution or by large quantities of fresh water. Eyes should be washed out with fresh water and then washed again with a boracic solution.

Alkaline and lead acid batteries should not be kept in the same compartment. Tools used for working on one type of battery should be thoroughly cleaned before being used on the other type.

Permit-to-work system

Permit-to-work forms should be used for any jobs which might be hazardous. The form is a document which states the work to be done and the safety precautions which must be adhered to when carrying out the task. It is a method whereby safety instructions are written down and transmitted to those entrusted with particular jobs. Much thought should go into the preparation of such permits, a predetermined safe drill should be formulated, all foreseeable hazards should be considered, and the appropriate precautions should be written down in a correct sequence.

Permits are not required for all jobs and it is essential that the system does not become overcomplicated. The permit should contain a carefully planned checklist to identify and eliminate or control hazards, plus arrangements for emergency procedures should an accident occur. Examples of jobs which require permits are:

(a) work on electrical equipment
(b) work on remote control machinery
(c) working aloft or outboard

A particular permit should only be issued by an officer who has experience in the appropriate work operation. The officer must ensure that the checks have been properly carried out and he should sign the permit only when he is satisfied that it is safe for work to proceed.

Entry into dangerous spaces

Despite the fact that much publicity has been given to the dangers of entering enclosed spaces there has been a long succession of tragedies over recent years. M910 should be studied in full as it gives case histories of some accidents. The notice also emphasizes the following points:

1 The atmosphere in any enclosed space may be incapable of supporting life due to a deficiency of oxygen or to the presence of toxic or flammable gases.
2 An unsafe atmosphere may be present in spaces such as cargo holds, ballast tanks, fresh water tanks, cofferdams, duct keels, etc.
3 A permit-to-work or similar scheme should be in operation before any enclosed spaces are entered.
4 Anyone who attempts to carry out a rescue without following correct procedures is endangering his own life and that of the person he is attempting to rescue.

This section should be read in conjunction with the Merchant Shipping Entry into Dangerous Spaces Regulations which are discussed in Chapter 2.

Most seamen are aware of the dangers associated with toxic and flammable gases. Inhalation of some toxic vapours can cause damage to the nervous system, the lungs, and to other vital organs, as well as causing brain damage and death. Hydrocarbon or flammable gas mixtures cause narcosis, which is a state of stupor, insensibility or unconsciousness. The symptoms of narcosis show first as eye irritation and headache, then diminished responsibility and a sense of dizziness which is described as being similar to drunkenness. If these warnings are ignored the result may be paralysis, insensibility and death.

Oxygen deficiency is possibly the most dangerous hazard within enclosed spaces as many seamen are unaware of the effects of such an atmosphere. The oxygen content of air is 21 percent. If the level falls to approximately 17 percent the atmosphere is unsafe and personnel will suffer impairment. Entry into an atmosphere of less than 10 percent oxygen causes unconsciousness, and death can occur if the victim is not quickly removed to the open air and resuscitated. Exposure to an atmosphere containing a low level of oxygen for only a few minutes can cause irreversible brain damage.

Many deaths have occurred in enclosed spaces on ships carrying what are generally classed as non-hazardous cargoes. Oxygen deficiency can be attributed to grain, timber, vegetable oils, steel, pig iron and many general cargoes. Oxygen can also be removed from the atmosphere in enclosed spaces by chemical reactions such as rusting or the hardening of paints and by the ingress of gases such as nitrogen or inert gas. Thus, all atmospheres in enclosed spaces must be considered as suspect and appropriate tests must be carried out using portable instruments.

The presence and the proportion of hydrocarbon gas in air is detected by the use of an 'explosimeter' or combustible gas indicator. This is a battery-operated instrument with an attached rubber sampling tube which is inserted into a compartment. The atmosphere within the space is drawn through the explosimeter by use of an aspirator bulb and a reading of gas percentage is obtained. The instrument should give a zero reading before entry is permitted. Contrary to the belief of some seamen, the explosimeter does not indicate the presence of toxic gases or oxygen deficiency.

Chemical absorption detectors must be used whenever the presence of toxic gases is suspected. One type consists of a hand bellows, a sampling tube, and a variety of glass tubes containing chemicals. Air is drawn through the chemical tube and the presence of gas is indicated by discolouration of the chemical. Each tube is only capable of detecting a certain gas and it may be necessary to test for several gases before entry is permitted.

Oxygen analysers should be carried on all ships and several types are available. All are capable of reaching remote corners by the use of sampling tubes. Aspiration draws the air through the analyser and the oxygen percentage content of the air is obtained. The reading must be 21 percent oxygen before entry into the space is permitted.

Tests should be taken by the remote sampling tube at several levels throughout the space and in several corners. The instruments should be carefully calibrated in fresh air before use and the manufacturer's instructions should be rigidly complied with. Any limitations on the use of an instrument should be understood.

If the above tests indicate that it is safe to enter a space, further tests within that space should be made by persons wearing breathing apparatus. Small portable explosimeters can be obtained for persons working in tanks which have been used for the carriage of hydrocarbon oils. These give audible warning if a build-up of gases occurs.

An efficient communication system should be set up by those working within the compartment. Thus all people within the space should be visually and audibly in contact with each other and with a stand-by man who must be posted at the compartment entrance. The stand-by man must be in constant attendance and he must be fully briefed on his actions in the event of an emergency. All the points in the following sample permit-to-work should be noted.

Sample permit-to-work

Entry into Enclosed Spaces
1 Date.
2 Period of validity (the period should not exceed 24 hours).
3 Location of work area.
4 Total number of persons entering the space.
5 The names of the designated crew members.
6 Information regarding the work to be carried out.
7 Time of testing the oxygen analyser.
  The oxygen content of the enclosed space.
8 Time of testing the hydrocarbon meter.
  Percentage of hydrocarbon gas in the space.
9 Time of testing the toxic gas detector.
  The results of toxic gas test.
10 Were the tests made at several levels or at different positions?
11 Time of entry and time of exit.
12 Is the space being continuously ventilated?
13 Is there a constant atmosphere monitoring system?
14 Is the agreed communication system functioning?
15 If VHF walkie-talkie radios are being used:
  (a) are they intrinsically safe (i.e. explosion proof)?
  (b) have they been tested?
16 Is there a stand-by man at the entrance?
17 Does the stand-by man understand his emergency procedure in the event of an accident?
18 Is the entrance clear?
19 If applicable, have the doors been secured?
20 Have warning notices been posted?
21 Is there adequate explosion-proof illumination?
22 Have precautions been taken to prevent entry of injurious substances into the space?
23 Have potential hazards been identified?
24 Have the bridge and engine room watchkeepers been informed?
25 Is protective clothing being worn?
26 Is all the equipment to be used of an approved type?
27 If repairs are to be carried out on machinery, is such machinery isolated from sources of power or heat?
28 Is fire-fighting equipment available?

29 Are the following items of equipment available and capable of functioning
   properly;
   Self-contained breathing apparatus
   Resuscitator
   Lifelines
   Safety harness
   Hoisting gear
   Stretcher
   Gas-tight (explosion-proof) torches
   Explosion-proof portable emergency lights
   First aid kit
   Spare torch and radio batteries
   Suitable fire extinguishers
30 Will the first men in be wearing breathing apparatus?

The emergency instructions should be printed at the bottom of the permit
and the permit should be signed by a responsible officer when he is satisfied
that all the safety procedures have been carried out.

Damage control

This is an aspect of seamanship which is sometimes neglected. Ideally,
damage control should be incorporated with the Emergency System. If
suitable, the Emergency Headquarters should also be used as the 'Damage
Control Station'.

The Station should have a stability calculator and a file containing
pre-calculated conditions of trims which might be encountered during
emergency situations. The conditions would naturally vary from ship to ship
but the following stability problems could be considered for light and loaded
voyages:

1 Flooding of fore peak or after peak.
2 Flooding of engine room.
3 Flooding of holds or cargo tanks.
4 Effect of flooding in areas isolated by watertight doors.
5 Stress on bulkheads due to adjacent compartments being flooded.
6 Effects of stress or strain due to grounding forward, aft, or amidships.
7 The use of ballast to change trim or to counteract listing in emergency
   situations.

Plans or diagrams which might be of use for damage control should be displayed in the Station or should be readily available. The following plans should be included:

1 Areas of the ship which are isolated by fire doors or fire-resistant bulkheads.
2 Remote controls for pumps, fans, sea valves, etc.
3 Watertight doors.
4 Fire-fighting and life-saving appliances.
5 Pipe line and pumps.
6 Trim and Stability Particulars Book.
7 Cargo.

The location of stores which might be required for damage repairs should be indicated. Such stores should include timber, cement, welding equipment, spare steel plates, portable pumps, and tools such as axes, crowbars, saws and hammers.

Damage control drills should be practised by the Emergency Team and the members should be instructed in their various roles in the event of:

> grounding
> collision
> breakdown of steering gear
> a derrick being dropped
> deck cargo shifting
> bulkhead fractures
> loss of an anchor
> ventilators and deck fittings being carried away

Training sessions should be held for instructing the team in:

> methods of constructing collison mats
> plugging shell plating holes above and below the waterline
> use of cement boxes
> shoring up bulkheads
> pumping out flooded compartments
> methods for towing and being towed

Suggestions for conducting the above operations in an efficient manner can be found in various seamanship books. Those methods should be adapted for use on individual ships. Damage control has its basis in forethought, training and predetermined knowledge.

A note for potential candidates for DoT Certificate of Proficiency/HND Nautical Science

Discuss safety awareness and methods for promoting safety awareness with interested crew members.

Devise a scheme for rescuing an unconscious man from an enclosed space on board your ship.

Use the above scheme to train crew members to deal with such an emergency and discuss its effectiveness.

Train crew members in the maintenance and use of the self-contained breathing apparatus and the air-hose breathing apparatus.

Devise a maintenance and inspection schedule for the fire equipment on board the ship.

Devise a maintenance and inspection schedule for the life-saving appliances on board the ship.

Compare your schedules with the company's maintenance scheme.

Discuss with other personnel the duties of a Safety Officer and a safety representative.

Consider how the effectiveness of the safety committee can be improved.

Further reading

Books

DANTON, G. *The Theory and Practice of Seamanship* (Routledge and Kegan Paul: London, 1978).

LEE, E. C. B. and LEE, K. *Safety and Survival at Sea* (Norton: New York and London, 1980).

ROBERTSON, D. *Sea Survival* (Elek: London, 1975).

RUSHBROOK, F. *Fire Aboard* (Brown Son and Ferguson: Glasgow, 1979).

RUTHERFORD, D. *Ship Safety Personnel: Role and Duties* (Griffin: London, 1982).

WRIGHT, C. H. *Survival at Sea* (Laver: Liverpool, 1977).

Manuals

Merchant Ship Search and Rescue Manual (IMO: London).

Marine Fire Prevention, Firefighting and Fire Safety (Brady: Bowie, Maryland).

Investigation reports

m.v. Burtonia, Court Report Number 8062 (HMSO: London, 1974).

m.v. Festivity, Court Report Number 8060 (HMSO: London, 1974).

m.v. Lovat, Court Report Number 8066 (HMSO: London, 1977).

Handbook

In Peril on the Sea? (MNAOA Handbook for Safety Representatives).

Journal

Joy, D. B. 'Fire Down Below—The "Hudson Transport" Case', *Seaways*, November 1983, pp 7–9.

Lavery, H. I. 'The Implementation of the "Safety Officials Act" for the Ship's Officer', *Seaways*, April 1983, pp 21–22.

Lavery, H. I. 'The 1986 Solas Training Requirements—An Impossible Ideal?' *Seaways*, August 1987, pp 8–10.

Rose, J. M. 'M Notices: The Helping Hands to Legislation?' *Seaways*, June 1987, pp 10–11.

2

Safety: Legislation

A   International Maritime Organization (IMO)

Before 22 May 1982 the International Maritime Organization was known as the Inter-Governmental Maritime Consultative Organization (IMCO). The Organization is based in London and the governing body, the Assembly, which consists of 128 Member States and one Associate Member, meets once every two years. A Council, consisting of 32 Member Governments elected by the Assembly, acts as IMO's governing body. IMO is a technical organization and most of its work is carried out by committees, e.g. the Maritime Safety Committee (MSC) which has 10 sub-committees such as Carriage of Dangerous Goods. The Marine Environment Protection Committee (MEPC) was established by the Assembly in 1973 and it has the responsibility for co-ordinating the Organization's activities in the prevention and control of pollution of the marine environment from ships.

The Secretariat, which has a staff of approximately 270 international civil servants, is headed by the Secretary General who is appointed by the Council. The objectives and activities of IMO are:

> To provide machinery for co-operation among Governments in the field of governmental regulations and practices relating to technical matters of all kinds affecting shipping engaged in international trade.

and

> To encourage the general adoption of the highest practicable standards in matters concerning maritime safety, efficiency of navigation, the prevention and control of marine pollution from ships and related legal matters.

To meet the objectives IMO has, within a twenty-five year period promoted the adoption of 30 conventions and protocols and over 600 codes and recommendations. Perhaps the four conventions which have the greatest practical application for mariners and shore staff are those which relate to Safety of Life at Sea, Regulations for Preventing Collisions, Prevention of Pollution and Load Lines. IMO's codes and recommendations cover a wide range of subjects, such as the carriage of particular types of cargo and the construction of specialized ships. A very brief synopsis of the contents of some of the conventions can be found in this chapter and the codes which have the most practical shipboard application will be referred to in the appropriate section of the book.

B   Safety of Life at Sea 1974 (SOLAS 74)

The International Convention for the Safety of Life at Sea, 1974, entered into force on 25 May 1980. The convention set out the minimum standards for the safe construction of ships and for the safety equipment which must be carried on board.

The 1978 Protocol to SOLAS 74

The Protocol, which deals mainly with tanker safety, entered into force on 1 May 1981. The SOLAS inert gas requirements for certain tankers were extended to all but the smallest tankers and new steering gear requirements were specified for certain sizes of tankers.

A new 'Steering Gear-Testing and Drills' regulation applied to all ships and the SOLAS requirements for inspection and certification of ships were made more stringent.

The 1981 SOLAS amendments

The 1981 SOLAS amendments and the 1981 Protocol amendments entered into force on 1 September 1984.

The 1983 SOLAS amendments

The 1983 SOLAS amendments entered into force on 1 July 1986. These amendments affected several chapters but the basic aim was to increase certain minimum standards in the following areas:

II-1 Floodable length, permeability, permissible length of compartments
II-2 Fire protection, detection and extinction
III Life-saving applicances

Chapter III was re-written and includes new provisions for lifeboats and liferafts, safer operational procedures for survival craft and rescue boats, better protection of survivors from environmental hazards and additional radio life-saving appliances.

The International Convention for the Safety of Life at Sea

The *Consolidated Text* of the above Convention, Protocol and Amendments was first published in 1986. Mariners should use the *Consolidated Text* for all practical requirements and it is recommended that this excellent volume should be carried on all vessels. The text contains eight chapters:

    I General provisions
 II-1 Construction—Subdivision and stability, machinery and electrical
      installations
 II-2 Construction—Fire protection, fire detection and fire extinction
  III Life-saving appliances and arrangements
   IV Radiotelegraphy and radiotelephony
    V Safety of navigation
   VI Carriage of grain
  VII Carriage of dangerous goods
 VIII Nuclear ships

Certification is dealt with in the Appendix and three annexes give a summary of the requirements relating to existing ships and future amendments to the 1974 SOLAS Convention.

The *Consolidated Text* contains 439 pages: a synopsis of the requirements would be impracticable, but attention is drawn to the following areas which affect some of the routine operations carried out on board ship. The *Consolidated Text* generally applies only to ships engaged on international voyages and the term 'new ship' means a ship, the keel of which was laid on or after 1 July 1986.

SOLAS Chapter II

Steering Gear—Requirements and Drills

The steering gear construction requirements are to be found in II-1, Regulation 29, and the operation, testing and drills requirements are located in V, Regulations 19-1 and 19-2.

Every ship must be provided with a main steering gear and an auxiliary steering gear and they shall be so arranged that the failure of one will not cause the other to be inoperative.

A specification that should be noted is that whilst the main steering gear must be capable of putting the rudder over from 35° on one side to 35° on the other at maximum draught and service speed (from 35° on either side to 30° on the other side in not more than 28 seconds) the auxiliary gear must only be capable of putting the rudder over from 15° to 15° in not more than 60 seconds at maximum draught but at one half of the maximum service speed or 7 knots (whichever is the greater).

Both steering gears must be capable of being operated from the navigation bridge and arranged to restart automatically after a power failure.

In every tanker (which includes chemical tankers and gas carriers) of 10,000 tons gross and upwards the main steering gear must consist of two or more identical power units both capable of operating the rudder independently.

Ships that have steering gear power units which are capable of simultaneous operation shall have more than one of the units operating when in areas where navigation demands special caution.

Emergency steering gear drills must be carried out at least once every three months to ensure that emergency steering procedures are practised. However, within 12 hours before a departure the following equipment must be checked and tested:

1 The main steering gear.
2 The auxiliary steering gear.
3 The remote steering gear control systems.
4 The steering positions located on the navigation bridge.
5 The emergency power supply.
6 The rudder angle indicators in relation to the actual position of the rudder.
7 The remote steering gear control system power failure alarms.
8 The steering gear power unit failure alarms;
9 Automatic isolating arrangements and other automatic equipment.

The full movement of the rudder, the steering gear and the connecting linkage should be visually inspected. Communications between the bridge and steering compartment should be checked.

All officers concerned with the operation or maintenance of the gear should be conversant with changeover procedures and block diagrams of such procedures, plus the operating instructions, should be permanently displayed on the bridge and in the steering compartment.

The dates of checks and tests and the dates and details of emergency steering drills should be recorded in a log book.

Emergency Source of Electrical Power in Cargo Ships

This information, which applies to new ships, is found in II-1, Regulation 43. Each ship must have a self-contained emergency source of electrical power sufficient to supply all the services that are essential for safety in an emergency. The following are the minimum services that must be provided.

1 Emergency lighting at every muster and embarkation station and over the side for a period of three hours
2 Emergency lighting for a period of 18 hours in
    all alleyways, stairs, exits and lifts
    the machinery spaces and generating stations
    all control stations and switchboards
    all stowage positions for firemen's outfits
    the steering gear compartment
    certain locations where fire and other pumps are located
3 Power for a period of 18 hours for
    the navigation lights and other lights required by the Collision Regulations
    the radiotelegraph main transmitter and main receiver
    the radiotelephone transmitter and receiver
4 Power for a period of 18 hours for
    all internal communication equipment required in an emergency
    navigational aids such as compasses, radars and automatic plotting equipment, echo-sounding device, speed and distance device and certain indicators
    the fire detection and fire alarm system
    the daylight signalling lamp, whistle, call points and internal signals
    one of the fire pumps if dependent upon the emergency generator for its source of power

In certain cases the emergency source shall also supply power to the steering gear for at least 30 minutes of continuous operation on ships of 10,000 tons gross tonnage and upwards and in any other ship for at least 10 minutes.

This regulation is complex and much detail has been omitted from this synopsis. The power source may be either a generator or an accumulator battery and it must be located above the uppermost continuous deck, readily accessible from the open deck and so placed that a fire at the main source of electrical power will not interfere with the operation of the emergency source.

Fireman's Outfit

Chapter II-2, Regulation 17. All ships shall carry at least two fireman's outfits, tankers shall carry two additional outfits and passenger ships carry additional outfits depending upon the aggregate lengths of passenger and service spaces.

The contents of a fireman's outfit can be divided into two basic equipment groups.

(a) Personal equipment which consists of:
    1 Heat and water resistant protective clothing.
    2 Boots and gloves of electrically nonconducting material.
    3 A rigid helmet.
    4 A hand held electric safety lamp which shines for at least three hours.
    5 An approved axe.
(b) An approved breathing apparatus which may be either:
    1 A smoke helmet or smoke mask with a suitable air pump and a length of hose able to reach from the open deck to any part of the holds or machinery spaces. If, in order to do so, the hose would exceed 36 metres in length, a self-contained breathing apparatus should be substituted or provided in addition; or
    2 A self-contained compressed-air-operated breathing apparatus with an air volume of 1,200 litres in the cylinders, or other self-contained breathing apparatus capable of functioning for at least 30 minutes.
       Suitable spare charges must be carried to the satisfaction of the administration (In IMO Regulations the term 'administration' means the government of the state whose flag the ship is entitled to fly).

Each breathing apparatus must be provided with a fireproof lifeline of sufficient length and strength with suitable means of attachment to the apparatus harness or to a separate belt.

All the equipment must be stored so as to be easily accessible and located in widely separated positions.

SOLAS—Chapter III

The most sweeping change to the 1974 SOLAS Convention has been the complete rewrite of this chapter. A study of the chapter shows that a particular emphasis has been placed on dry abandonment into survival craft and on the training for use, and the maintenance of, survival equipment. This chapter applies in full to new ships from 1 July 1986 and in the case of existing ships partly from that date but mainly from 1 July 1991. Marine personnel, both ashore and afloat, should therefore closely study this chapter to ascertain which particular regulations apply to particular ships. As radical changes have been made to the requirements of some major items of survival equipment and to training procedures, some of the changes are noted below. The term 'survival craft' means 'a craft capable of sustaining the lives of persons in distress from the time of abandoning the ship'.

Survival Craft Emergency Position—Indicating Radio Beacons

Chapter III, Regulation 6, Section 2.3. This regulation applies to all ships but existing ships do not have to comply until 1 July 1991. One manually activated emergency position-indicating radio beacon must be carried on each side of the ship. They must be stowed in a manner whereby they can rapidly be placed in survival craft.

Two-way Radiotelephone Apparatus

Chapter III, Regulation 6, Section 2.4. This regulation applies to all ships but existing ships do not have to comply until 1 July 1991. Two-way radio-telephones must be provided for communication between survival craft, a minimum of three must be provided on each ship. The radiotelephone appara-tus which is used on board many ships for operational procedures, such as berthing, may be used for survival craft purposes if such apparatus complies with Regulation 14.3 of Chapter IV.

Muster Lists and Emergency Instructions

Chapter III, Regulations 8 and 53. These regulations apply to all ships. Clear instructions which are to be followed in the event of an emergency must be provided for every person on board and muster lists should be exhibited in conspicuous places throughout the ship (including the bridge, engine-room and accommodation spaces).

A muster list must give the details of the general emergency alarm signal (seven or more short blasts followed by one long blast) and the action to be taken when the alarm is sounded. Emphasis should be put on how the *actual* order to *abandon ship* will be given. The muster list should show the duties assigned to the different crew members and the following duties should be included:

1 The closing of the watertight doors, fire doors, valves, scuppers, side-scuttles, skylights, portholes and any similar opening.
2 Putting equipment into survival craft.
3 Preparing and launching survival craft.
4 General preparation of other life-saving appliances.
5 The muster of passengers (this should include wives).
6 Use of communication equipment.
7 The manning of fire parties assigned to deal with fires.
8 Special duties assigned in respect of the use of fire-fighting equipment and installations.

Many ships now use emergency teams which are trained to deal with specific emergencies and many administrations encourage the use of such a system.

Emergency team duties, therefore, will have to be carefully stipulated, co-ordinated and integrated with the obligatory muster list.

In addition to the above, the muster list shall state which officers have the responsibility for ensuring the maintenance and ready availability of life-saving and fire appliances and it shall specify substitutes for key persons who may become disabled in emergencies (one hopes that such substitutes *will* be available now that many ships have reduced manning scales). The muster list must be prepared before a ship proceeds to sea and amended as necessary.

Additional duties are stipulated for crew members on passenger ships and the format of the list used on such ships must be approved.

Survival Craft Operating Instructions

Chapter III, Regulation 9, applies to all ships.

Posters or signs must be exhibited on, or in the vicinity of, survival craft and their launching controls. The posters or signs must:

1 Illustrate the purpose of controls and the procedures for operating the appliance and give relevant instructions or warnings.
2 Be easily seen under emergency lighting conditions.
3 Use IMO approved symbols.

Manning of Survival Craft and Supervision

Chapter III, Regulation 10, applies to all ships.

Many mariners may believe that reduced manning levels will make the implementation of this regulation difficult. However, the regulation is unambiguous and the following requirements should be complied with:

1 There must be a sufficient number of trained persons for mustering and assisting untrained persons.
2 There must be a sufficient number of crew members for operating the survival craft and launching arrangements for abandonment by all persons on board.
3 A deck officer or certificated person must be placed in charge of each survival craft which is used (however, persons practised in the handling and operation of liferafts may be placed in charge of liferafts), and a second-in-command must be nominated in the case of lifeboats.
4 The person in charge of the survival craft must have a list of the survival craft crew and must ensure that the crew under his command are acquainted with their duties; in lifeboats the second-in-command must also have a list of the lifeboat crew.
5 If a passenger ship lifeboat has a radiotelegraph installation, a person capable of operating the equipment must be assigned to the lifeboat.

6 Each motorized survival craft must have a person capable of operating the engine and carrying out minor adjustments.

7 The Master must ensure the equitable distribution of trained personnel among the ship's survival craft.

It may be advisable, therefore, that the Safety Officer of a ship whose crew is not proficient in abandonment procedures initiates a comprehensive programme of safety training for the personnel.

Abandon Ship Training and Drills, and the Training Manual

Chapter III, Regulation 18, applies to all ships.

A training manual which complies with Regulation 51, must be provided in each crew messroom and recreation room or in each crew cabin. The manual must contain instructions and information on the life-saving appliances and on the best methods of survival. The manual should be written in easily understood terms and illustrated wherever possible. Audio-visual aids may be considered as part of the manual. The following must be explained in detail:

1 Donning of lifejackets and immersion suits.
2 Muster at the assigned stations.
3 Boarding, launching and clearing the survival craft and rescue boats.
4 Method of launching from within the survival craft.
5 Release from launching appliances.
6 Methods and use of devices for protection in launching areas where appropriate.
7 Launching area illumination.
8 Use of all survival equipment.
9 Use of all detection equipment.
10 Use of radio life-saving appliances (with the assistance of illustrations).
11 Use of drogues.
12 Use of engines and accessories.
13 Recovery of survival craft and rescue boats, including stowage and securing.
14 Hazards of exposure and the need for warm clothing.
15 Best use of survival craft facilities in order to survive.
16 Methods of retrieval, including the use of helicopter rescue gear, breeches buoy and shore life-saving apparatus and ship's line-throwing apparatus.
17 All other functions contained in the muster list and emergency instructions.
18 Instructions for emergency repair of the life-saving appliances.

As many of the above procedures are contained within the syllabuses of 'Survival at Sea' courses at certain nautical colleges, I would recommend that all officers be encouraged to attend such courses.

Regulation 18 requires that each crew member shall participate in at least one abandon ship drill and one fire drill every month. The drills must take place within 24 hours of the ship leaving a port if more than 25% of the crew have not taken part in such drills on board that particular ship within the previous month. There are additional requirements for the mustering of passengers on passenger ships.

Each abandon ship drill must include the following:

1 The summoning of passengers and crew to the muster stations by the sounding of the general alarm.
2 Ensuring that everyone understands the abandon ship order as specified in the muster list.
3 Reporting to stations and preparing for the duties described in the muster list.
4 Checking that everyone is suitably dressed.
5 Checking that lifejackets have been put on properly.
6 Preparing and lowering at least one lifeboat.
7 Starting and operating the lifeboat engine.
8 The operation of davits used for launching liferafts.

Different lifeboats should be lowered at successive drills and the drills should be conducted, as far as practicable, as if there is an actual emergency. Each lifeboat with its assigned operating crew should be manoeuvred in the water at least once every 3 months during an abandon ship drill. Ships on short international voyages may meet different criteria.

Rescue boats which are not lifeboats should be manoeuvred in the water by the assigned crew at least every 3 months, but in general this should be carried out each month.

Because lifeboat and rescue boat launching drills carried out with the ship making headway involve dangers, such drills should only be practised in sheltered waters under the supervision of an officer experienced in such drills.

The emergency lighting for mustering and abandonment must be tested at each abandon ship drill.

The new emphasis on training is apparent in Regulation 18. Each crew member must be given training in the use of the life-saving appliances as soon as possible after joining but such training must be given not later than 2 weeks from joining. Instructions to the crew in the use of the life-saving appliances and in survival-at-sea must be given each month. Instructions may be given on different parts of the ship's life-saving system but the complete system must be covered within any period of 2 months. Each crew member must be given instructions on the following subjects:

1 Operation and use of the ship's inflatable liferafts.
2 Problems of hypothermia, first-aid treatment of hypothermia and other appropriate first-aid procedures.

3 Special instructions necessary for use of the ship's life-saving appliances in severe weather and severe sea conditions.

Special training must be given in the use of any davit-launched liferafts.

The administration (i.e. the government of the state whose flag the ship is entitled to fly) may require the following information to be recorded in a log-book:

> The date when musters are held
> Details of abandon ship drills
> Details of fire drills
> Drills of other life-saving appliances
> On-board training

If a full muster, drill or training session is not held at the appointed time, an entry must be made in the log-book stating the circumstances and the extent of the muster, drill or training session held.

Although it is not stipulated in this regulation, the Safety Officer may find that the contents of other regulations can be used as the basis for instruction and/or training periods. Such regulations include:

III-12 Launching Stations
III-13 Stowage of Survival Craft
III-14 Stowage of Rescue Boats
III-15 Survival Craft Launching and Recovery Arrangements
III-16 Rescue Boat Embarkation, Launching and Recovery Arrangements
III-48 Launching and Embarkation Appliances
III-49 Line-throwing Appliances

Operational Readiness, Maintenance and Inspections

Chapter III, Regulation 19, applies to all ships.

This regulation emphasizes the important point that survival equipment is of little use if it is not ready for use at all times, as it states that: 'Before the ship leaves port and at all times during the voyage, all life-saving appliances shall be in working order and ready for immediate use'.

Instructions for on-board maintenance of life-saving appliances shall include the following for each appliance:

1 A checklist to be used for a monthly inspection of all life-saving appliances, including lifeboat equipment, to ensure that such appliances are complete and in good order.
2 Maintenance and repair instructions.
3 Schedule of periodic maintenance.
4 Diagram of lubrication points with the recommended lubricants.
5 List of replacable parts.

6 List of sources of spare parts.
7 Log for records of inspections and maintenance.

A report of the monthly inspection must be entered in the ship's log-book.

A shipboard planned maintenance schedule may incorporate the instructions listed above.

Falls used in launching shall be turned end-for-end at intervals of not more than 30 months and be renewed when necessary due to deterioration or at intervals of not more than 5 years, whichever is the earlier. Adequate spares for the appliances must be carried.

The following tests and inspections shall be carried out weekly:

1 All survival craft, rescue boats and launching appliances to be visually inspected to ensure that they are ready for use.
2 All engines in lifeboats and rescue boats to be run ahead and astern for a total period of not less than 3 minutes (in special cases this may be waived for ships constructed before 1 July 1986).
3 The general emergency alarm system to be tested.

Every inflatable liferaft and inflatable lifejacket shall be serviced at intervals not exceeding 12 months at an approved servicing station (in certain cases this may be extended to 17 months). All repairs and maintenance of inflated rescue boats shall be carried out in accordance with the manufacturer's instructions and only emergency repairs may be carried out on board, permanent repairs shall be carried out at an approved servicing station. Hydrostatic release units shall be serviced at intervals not exceeding 12 months at a competent servicing station (in certain cases this may be extended to 17 months).

Inflatable Liferafts

Chapter III, Regulation 39 (Inflatable Liferafts) and Regulation 38 (General Requirements for Liferafts) apply to all ships.

The regulations are very detailed and should be closely studied but attention should be paid to the following points:

1 The liferaft shall have an efficient painter and the length must be not less than twice the distance from the stowed position to the waterline in the highest sea-going condition or 15 metres, whichever is the greater.
2 The liferaft painter system shall provide a connection between the ship and the liferaft and shall be so arranged as to ensure that the liferaft when released and, in the case of an inflatable liferaft, inflated is not dragged under by the sinking ship, i.e. there must be a float-free arrangement.
3 If a weak link is used in the float-free arrangment, it shall not be broken by the force required to pull the painter from the container and, if applicable, shall be strong enough to permit the inflation of the liferaft.

4 If a hydrostatic release unit is used in the float-free arrangement, it shall automatically release the liferaft at a depth of not more than 4 metres.
5 The main buoyancy chamber shall be divided into not less than two separate compartments.
6 The floor shall be capable of being insulated against cold (usually by a double floor which inflates automatically but which can be deflated and reinflated by the occupants).
7 The gas used for inflating the raft shall be non-toxic,
8 At least one entrance shall be fitted with a semi-rigid boarding ramp.
9 Entrances not fitted with a boarding ramp shall have a boarding ladder.
10 There shall be means inside the liferaft to assist persons to pull themselves into the liferaft from the ladder.
11 If the liferaft becomes inverted it shall be capable of being righted in a seaway in calm weather by one person.
12 A manually controlled lamp visible at night for a distance of at least 2 miles for not less than 12 hours shall be fitted to the top of the canopy powered by a sea-activated or a dry chemical cell and which shall light automatically when the liferaft inflates.
13 Another manually controlled lamp shall be fitted inside the liferaft capable of continuous operation for a period of at least 12 hours and which shall light automatically when the liferaft inflates.

A liferaft should be packed in a suitable container in such a way as to ensure as far as possible that the liferaft inflates in an upright position. A container should be marked with the following information:

Maker's name or trade mark
Serial number
Name of approved authority and the number of persons it is permitted to carry
SOLAS
Type of emergency pack enclosed
Date when last serviced
Length of painter
Maximum permitted height of stowage above the waterline (this relates to drop-test height—at least 18 metres but in some cases higher—and painter length)
Launching instructions

Similar information must be marked on the liferaft itself.
The type of emergency pack will generally be 'SOLAS A' pack (passenger ships on short international voyages may have liferafts equipped with 'SOLAS B' packs). Mariners should carefully study Regulation 38 to ensure that a comprehensive knowledge of liferaft equipment is gained (it is too late to

attempt to acquire knowledge during emergency situations), but it should be noted that liferafts now have:

> four rocket parachute flares
> six hand flares
> two buoyant smoke signals
> thermal/protective aids sufficient for 10% of the liferaft complement or two, whichever is the greater
> an efficient radar reflector
> two sea-anchors of a greatly improved type
> seasickness bags (which will probably be very necessary in most sea conditions)

It should be noted that Regulation 26 stipulates that liferaft capacity should be for 100% of the ship's complement instead of 50% as before. Existing ships have a 'period of grace' until 1 July 1991 before compliance with this standard is officially required.

Lifeboats

Chapter III, Regulation 41, deals with the general requirements for lifeboats and Regulations 42 to 46 inclusive deal with the permitted 'sub-species' of lifeboats. Totally enclosed lifeboats, which must comply with Regulation 44, are required on new cargo ships in place of the traditional open lifeboat. In general, the open lifeboat will gradually disappear by 1 July 1991. Once again the Regulations should be studied in full but mariners should note that all lifeboats should be of sufficient strength to enable them to be safely lowered into the water when fully loaded and should be capable of being launched and towed when the ship is making headway at a speed of 5 knots in calm water. Other pertinent features are:

1 Every cargo ship lifeboat to be so arranged that it can be boarded by its full complement of persons in not more than three minutes from the time the instruction to board is given.
2 A boarding ladder to be provided that can be used on either side of the lifeboat to enable persons in the water to board, the lowest step of the ladder to be not less than 0.4 metres below the lifeboat's light waterline.
3 The lifeboat to be so arranged that helpless people can be brought on board either from the sea or on stretchers.
4 All surfaces on which persons might walk to have a non-skid finish.
5 Every lifeboat to be powered by a compression ignition engine with either a manual starting system or a power starting system with two independent rechargeable energy sources, both systems to be capable of starting the engine at an ambient temperature of $-15°C$ within 2 minutes (unless otherwise permitted by the administration).

6 Every lifeboat engine to be capable of being operated for not less than 5 minutes after starting from cold with the lifeboat out of the water and to be capable of operation when the lifeboat is flooded up to the crankshaft.

7 Lifeboat speed to be at least 6 knots with sufficient fuel to run for a period of not less than 24 hours.

8 The engine arrangements to be enclosed in a fire-retardant casing.

9 Means to be provided for recharging all engine-starting, radio and search-light batteries.

10 Water-resistant instructions for starting and operating the engine to be mounted in a conspicuous place near the engine mounting.

11 Each lifeboat to have at least one drain valve which shall automatically open to drain water from the hull when the lifeboat is not waterborne and which shall automatically close when the vessel is waterborne, the position of the drain valve to be clearly indicated.

12 Each lifeboat to be fitted with a release device to enable the forward painter to be released when under tension.

13 A manually controlled light visible on a dark night for not less than 12 hours to be fitted to the top of the cover or enclosure.

14 A lamp which provides illumination for not less than 12 hours to be fitted inside the lifeboat (an oil lamp is not permitted for this purpose).

The carrying capacity of a lifeboat is calculated by using either the number of persons wearing lifejackets that can be seated in a normal position without interfering with the operation of the lifeboat or standard dimensions for seated personnel. Every lifeboat that is launched by a fall or falls is to be fitted with a release mechanism complying with the following requirements:

1 The mechanism to be arranged so that all hooks are released simultaneously.

2 The mechanism to have two release capabilities, 'normal' when there is no load on the hooks, and 'on-load' when there is a load on the hooks, the latter capability to be adequately protected against accidental or premature release.

3 The release control to be clearly marked in a colour that contrasts with its surroundings.

4 The mechanism to be designed with a factor safety of 6.

Regulation 41 should be studied to ascertain the items of equipment that must be carried. As is the case with liferafts, some items such as a survival manual and a few thermal protective aids are additional to earlier regulations and other items such as painters and a sea-anchor are designed to higher standards.

A lifeboat must be marked as follows:

1 The dimensions and the number of persons which it carries to be marked in clear permanent characters.

2 The name and port of registry marked on each side of the bow in block roman capitals.

3 Means of identifying the ship to which the lifeboat belongs and the lifeboat's number to be marked so as to be visible from above.

Rescue Boats (Chapter III, Regulation 47)

A rescue boat is a boat designed to rescue persons in distress and to marshall survival craft (a survival craft is a craft capable of sustaining the lives of persons in distress from the time of abandoning the ship) and rescue boats are becoming a standard requirement on most vessels.

Rescue boats may be either of rigid or inflated construction or a combination of both. A rescue boat is not considered as one of the required number of survival craft but is additional to them unless the option given in Regulation 26 is taken whereby a lifeboat may be accepted as a rescue boat provided that it also complies with the requirements for a rescue boat.

A rescue boat must be capable of carrying at least 5 seated persons and one lying down. It must have a bow cover and be capable of manoeuvring at speeds of up to 6 knots and maintaining that speed for at least 4 hours.

Launching and Embarkation Appliances

Chapter III, Regulation 48: this regulation should be studied in conjunction with Regulation 15, Survival Craft Launching and Recovery Arrangements and Regulation 16, Rescue Boat Embarkation, Launching and Recovery Arrangements.

Launching appliances must be capable of being lowered against an adverse heel of 20° and a trim of 10°. However, in oil tankers, chemical carriers and gas carriers, with a final angle of heel greater than 20°, launching appliances must be capable of operating at the final angle of heel on the lower side of the ship.

A launching mechanism shall be arranged so that it may be actuated by one person, it must depend on gravity or stored mechanical power (i.e. launching power must be independent of the ship's power supplies) and it shall remain effective under conditions of icing.

Every rescue boat launching appliance shall be fitted with a powered winch motor of a capacity which will enable the rescue boat to be raised from the water with its full complement of persons and equipment. An efficient hand gear shall be provided for the recovery of each survival craft and rescue boat. Where davit arms are recovered by power, safety devices are to be fitted to cut off the power automatically before the arms reach the stops.

Every launching appliance is to be fitted with brakes capable of stopping and holding a fully loaded survival craft or rescue boat during launching. Manual brakes must be arranged so that the brake is always applied unless the operator holds the control in the 'off' position.

Falls shall be of rotation-resistant and corrosion-resistant steel wire rope.

Basic Life-saving Appliances on 'New' Cargo Ships

The regulations contained in SOLAS Chapter III are rather tortuous and complex with much cross-referencing between various regulations. Mariners should carefully check that the equipment on board ship complies with the various regulations: the following list is given as an indication of the life-saving appliances required on a 'typical' new cargo ship, i.e. a ship being constructed on or after 1 July 1986.

1 Totally enclosed, fire retardant, self-righting lifeboats sufficient for the total complement to be carried on both sides of the vessel (on chemical and gas tankers each lifeboat must have a self-contained air support system). Free-fall lifeboats carried aft may be an alternative. Most oil, chemical and gas tankers will have lifeboats fitted with a fire protection (external water spray) system.
2 A rescue boat (unless one of the lifeboats has been accepted as a rescue boat).
3 Liferafts sufficient for the total complement capable of being launched from either side or liferafts for the total complement on each side of the vessel. If the survival craft are stowed more than 100 metres from the stem or stern an additional liferaft is required to be stowed as far forward or aft as is practicable.
4 Lifebuoys as follows:

Length of ship (metres)	Minimum number
Under 100	8
100 and under 150	10
150 and under 200	12
200 and over	14

The lifebuoys must be distributed so as to be as readily available as practicable and at least one in the vicinity of the stern.

At least one lifebuoy on each side of the ship shall be fitted with a buoyant lifeline equal in length to not less than twice the height at which it is stowed above the waterline at any time or 30 metres, whichever is the greater.

Not less than half of the lifebuoys must have self-igniting lights, not less than two of which must be provided with self-activating smoke signals and which must be capable of quick release from the navigating bridge. Lifebuoys fitted with lights or smoke signals shall not be the lifebuoys provided with lines and shall be equally distributed to port and starboard. Self-igniting lights on tankers shall be of the electric battery type.

Each lifebuoy shall have the ship's name and port of registry marked on it in black roman capitals.

5 Lifejackets to be provided for every person on board the vessel plus a sufficient number for persons on watch and for use at remotely located survival craft stations.

6 Immersion suits to be provided for every person assigned to crew the rescue boat plus as deemed necessary under Regulation III 27.

7 One EPIRB on each side of the vessel.

8 At least three two-way radiotelephones.

9 Retro-reflective tape or material on all lifeboats, liferafts, lifebuoys and lifejackets.

10 A portable radio apparatus for survival craft.

11 Not less than twelve rocket parachute flares stowed on or near the navigating bridge.

12 An on-board communications and alarm system.

13 A line-throwing appliance with four projectiles and lines each capable of travelling a distance of 230 metres in calm weather with reasonable accuracy.

Features of a SOLAS lifejacket (non-inflatable)

Under Chapter III, Regulations 32 and 27, a non-inflatable lifejacket should meet the following standards:

1 Does not sustain burning or continue melting after being totally enveloped in a fire for a period of 2 seconds.

2 Capable of being correctly put on within 1 minute.

3 Capable of being worn inside out or is clearly capable of being worn in one way only and, as far as possible, cannot be put on incorrectly.

4 Comfortable to wear.

5 Allows the wearer to jump from a height of 4.5 metres into the water without injury and without dislodging or damaging the lifejacket.

6 Lifts the mouth of an exhausted or unconscious person not less than 120 mm clear of the water with the body inclined backwards.

7 Turns the body of an unconscious person from any position to one where the mouth is clear of the water in 5 seconds.

8 Buoyancy not reduced by more than 5% after 24 hours submersion in fresh water.

9 Enables the person wearing it to swim a short distance and to board a survival craft.

10 Fitted with a whistle firmly secured by a cord.

11 On cargo ships each lifejacket to be fitted with a light as under Regulation 32.3 (with respect to cargo ships constructed before 1 July 1986, this

feature shall apply not later than 1 July 1991). The light shines for at least 8 hours.

12 Fitted with retro-reflective material.

General requirements for life-saving appliances

Chapter III, Regulation 30, gives the following requirements for life-saving appliances:

1 Constructed with proper workmanship and materials.
2 Not damaged in stowage throughout the air temperature range of −30°C to +65°C.
3 If likely to be immersed in seawater during use, operate throughout the seawater temperature range of −1°C to +30°C.
4 Where applicable, rot-proof, corrosion-resistant, and not unduly affected by seawater, oil or fungal attack.
5 Resistant to sunlight deterioration.
6 Of a highly visible colour.
7 Fitted with retro-reflective material.
8 If to be used in a seaway, capable of satisfactory operation in a seaway.

The government of the state whose flag the ship is entitled to fly shall determine the period of acceptability of life-saving appliances which are subject to deterioration with age. Such life-saving appliances shall be marked with a means for determining their age or the date by which they must be replaced

Testing of life-saving appliances

Chapter III, Regulation 4, makes the requirements of the IMO publication *Recommendations on Testing and Evaluation of Life-Saving Appliances* mandatory for governments giving approval to life-saving appliances and arrangements.

C Prevention of pollution 1973 (MARPOL 73)

The International Convention for the Prevention of Pollution from Ships, 1973, came into force on 2 October 1983. This convention contains regulations which are designed to prevent pollution caused accidentally or during routine operations by ships transporting oil cargoes, by noxious or harmful cargoes, and by sewage and garbage. The requirements for the storing, treating, and discharging of such substances are set out and also the procedures for the reporting of spillages. The technical measures are stipulated in five annexes:

    I Prevention of Pollution by Oil

    II Control of Pollution by Noxious Liquid Substances in Bulk (e.g. chemicals).

    III Prevention of Pollution by Harmful Substances carried in Packages (e.g. packaged forms, containers, tanks).

    IV Prevention of Pollution by Sewage.

    V Prevention of Pollution by Garbage.

The 1978 Protocol to MARPOL 73

The international conference on 'Tanker Safety and Pollution Prevention' (TSPP) which was held in 1978, in addition to issuing the SOLAS Protocol, recognised 'the need to improve further the prevention and control of marine pollution from ships, particularly oil tankers'. The MARPOL 78 Protocol was therefore promulgated and it came into force on 2 October 1983.

The Protocol deals mainly with the requirements for tankers to comply with legislation concerning segregated ballast tanks (SBTs), the clean ballast tank system (CBT), and crude oil washing (COW).

The Protocol also makes strict provision for inspection and certification procedures to ensure that all ships comply with pollution prevention measures. All but the smallest vessels must undergo an initial survey before being issued with the 'International Oil Pollution Prevention Certificate' (IOPP Certificate). Periodical surveys at intervals not exceeding five years must be carried out, and also at least one intermediate survey half-way through the period of validity.

The original MARPOL 73 Convention and 78 Protocol are collectively known as MARPOL 73/78. Details of particular regulations from MARPOL 73/78 can be found in Chapter 6.

Regulations for the prevention of pollution by oil—Annex I, MARPOL 73/78, 1986 consolidated edition

IMO, which is to be commended for consolidating all of the current provisions of Annex I in a single publication as a series of new measures to prevent pollution by oil (the 1984 Amendments), came into force in 1986. Indeed, the necessary but continuous improvement in anti-pollution standards and technical measures has unfortunately made this area a legal minefield for ship and shore managers alike.

The new measures require that existing ships comply with regulations concerning the installation of oil discharge, monitoring and control systems and oily-water-separating and oil-filtering systems.

One of the main points of the 1984 Amendments deals with ships operating in special areas (basically the Mediterranean, Baltic, Black and Red Seas and the 'Gulf' area). Regulation 10 requires the installation of a device to oily-water-separating equipment which stops overboard discharge once the oil content of the mixtures exceeds 15 ppm. Any residues which cannot meet the 15 ppm standard must be retained on board the ship.

Regulation 20 changed the format of the Oil Record Book. Non-tankers now have to carry an Oil Record Book Part I (Machinery Space Operations) whilst tankers will additionally have to carry Part II (Cargo/Ballast Operations). The contents of the Oil Record Book are discussed later in this chapter within the context of the British Prevention of Oil Pollution Regulations 1983. Ships' masters should note that each completed page of the Oil Record Book must now be signed by the master.

Regulations for the control of pollution by noxious liquid substances in bulk—Annex II, MARPOL 73/78

Annex II is particularly complex and the effective date of this annex was delayed twice, the second delay being to allow the adoption of amendments to the annex. Thus, Annex II, which incoporates amendments, entered into force on 6 April 1987.

Annex II contains sixteen regulations; however, some confusion may be caused by the fact that two regulations are designated with both a number and the letter A. Thus, although the last regulation is number 14, regulations 5A and 12A bring the total number to sixteen.

All the regulations are important but the following are worth emphasizing:

Regulation 3—Categorization and listing of noxious liquid substances. Chemicals which are carried by sea and which could harm the marine environment are divided into four categories, 'A' being the most dangerous and 'D' the least. Appendices to the annex contain guidelines for the categorization of chemicals and the agreed categories of chemicals transported by sea.

Regulation 5—Discharge of noxious liquid substances. This is a particularly important regulation as it states the requirements for the discharge of substances and such requirements vary according to the categorization. The discharge into the sea of ballast water, tank washings, and other residues containing Category 'A' chemicals is completely banned. Residues which are the result of tank-washing operations of tanks which contained Category 'A' substances must be discharged into a shore reception facility but any water subsequently added to the tank may be discharged into the sea provided that:

the ship is travelling at a speed of at least seven knots
the discharge is below the waterline
the ship is not less than 12 miles from the shore *and*
the water is not less than 25 metres deep

Other requirements pertain to categories 'B' to 'D' and to operations in 'special areas' (Black Sea and Baltic Sea areas).

Regulation 5A—Pumping, piping and unloading arrangements. This is a new regulation which for the first time introduces certain parameters with regard to category 'B' and 'C' substances. To ensure that quantities of residues are not left in cargo tanks, every ship constructed on or after 1 July 1986 must have pumping and piping arrangements that ensure, by testing with water under favourable pumping conditions, that each tank designated for the carriage of 'B' and 'C' substances does not retain a quantity of residue in the tank's piping or around the suction in excess of certain stipulated quantities. Thus, the amount of category 'B' substances must not exceed 0.1 cubic metres and for category 'C' substances the upper limit is 0.3 metres.

Ships built before 1 July 1986 must comply with similar but slightly less stringent requirements.

Regulation 7—Reception facilities and cargo unloading terminal arrangements. This regulation is of particular help to ships' officers as it requires contracting parties to ensure that facilities are provided at shore installations for the reception of chemical wastes. IMO has published guidelines to assist governments of contracting parties to evaluate the adequacy of reception facilities in their ports.

Regulation 8—Measures of control. This regulation requires contracting parties to appoint surveyors to ensure that high operational standards are maintained. Operations must be recorded in the Cargo Record Book and in some cases the book must be endorsed by a surveyor. It should be noted that tanks which have contained category 'A' substances must now be washed *before* the ship leaves the unloading port. This operation of washing at a loading port is known as 'prewashing', a term which might lead to some confusion.

Prewashing is also required for certain 'B' and 'C' substances anywhere in the world but in special areas it is a requirement for all 'B' and 'C' cargoes. This regulation also gives details of exceptions to the prewash requirements.

Regulation 9—Cargo Record Book. All ships to which the Annex applies must carry this record book in which various operations, such as loading, discharging, tank washing and ballasting, are recorded. The book must be retained on board for at least three years after the last entry and it may be inspected by shore authorities to ensure that the requirements of Annex II are complied with.

Regulation 10—Surveys. The surveys which are required for chemical tankers are now similar to those required for oil tankers, i.e.
(a) an initial survey before the ship is put into service;
(b) periodical surveys;

(c) an intermediate survey during the period of validity of an 'International Pollution Prevention Certificate for the Carriage of Noxious Liquid Substances in Bulk' (the certificate lasts for a period of five years).

In a separate publication to that of Annex II, IMO has produced *Guidelines for Surveys under Annex II of MARPOL 73/78* which is intended to assist administrations in formulating procedures for carrying out the surveys which took effect on 6 April 1987. Shore and ship management should have knowledge of the document.

Other regulations stipulate which ships must have an 'International Pollution Prevention Certificate for the Carriage of Noxious Liquid Substances in Bulk'. It should be noted that chemical tankers which are surveyed under the provisions of the International Bulk Chemical Code or Bulk Chemcial Code (see Chapter 7) will be issued with a 'Certificate of Fitness' only.

Procedures and Arrangements Manual. IMO has published a document *Standards for Procedures and Arrangements for the Discharge of Noxious Liquid Substances,* which provides a uniform international basis for the shipboard *Procedures and Arrangements Manual.* The shipboard manual's main purpose is to enable the ship's officers to identify the arrangements and operational procedures for all the various aspects of the transportation of chemicals. Each manual contains much detailed information and should have a standard format A. The manual should be in two parts and the first part should be divided into four sections:

1 A description of the main features of Annex II
2 A description of the ship's equipment and arrangements
3 Cargo unloading procedures and tank stripping
4 Procedures relating to the cleaning of cargo tanks, residue discharge, ballasting and deballasting

The second part of the manual should only contain information pertaining to the ship on which the manual is carried. Information could include a table of noxious substances which the ship is certified to carry, a table identifying in which tanks each noxious liquid may be carried, descriptions of equipment such as cargo heating and temperature control systems and information of the tanks carrying noxious liquids.

It should be noted that the manual is an operational manual and not a safety guide.

MARPOL 73/78 Annexes III to V

These three annexes are optional and some countries which have accepted the 'parent' Convention have not yet adopted annexes III to V. However, Annex V, 'Prevention of Pollution by Garbage', was adopted by sufficient countries to enable it to enter into force on 31 December 1988.

The Annex applies to all ships and its purpose is to prevent the pollution which can be caused by the dumping of food, domestic and operational waste from ships conducting their normal day-to-day activities. It should be noted that there is a complete ban on the dumping into the sea of all plastics, including synthetic ropes, synthetic fishing nets and plastic garbage bags. Ships' staff should note that environmental pressure groups may, quite correctly, be keeping an even closer watch on the rubbish-handling procedures of short-sea ferries due to the implementation of Annex V.

Some types of garbage may be dumped if strict conditions are complied with. Dunnage, lining and packing materials can only be disposed of at sea when the vessel is more than 25 miles from land. Food wastes and all other garbage (which includes paper products, rags, bottles, crockery, glass and metal) cannot be dumped within 12 miles of land unless it has been passed through a grinder or comminuter. In any case, the minimum distance from land when dumping is permitted is 3 miles.

Extremely strict controls apply to the 'special areas' and some sea areas in the Middle East where garbage pollution is prevalent. In those areas dumping of all forms of garbage, except food wastes, is completely banned, and food wastes cannot be dumped into the sea within 12 miles of land.

Contracting parties to the Convention must provide facilities in port for the reception of garbage.

D   Other IMO Safety Initiatives

IMO search and rescue manual (IMOSAR manual)

This manual provides international guidelines for a common maritime search and rescue (SAR) policy. The intention of the manual is to encourage all coastal states to develop their rescue organizations on similar lines, thus enabling adjacent states to co-operate and provide mutual assistance. The manual is in two parts. Part I deals with the organization of existing services and facilities necessary to provide practical and economical SAR coverage of a given area. Part 2 contains information to assist all personnel participating in SAR operations and exercises. Details of the Maritime Search and Rescue Recognition Code (MAREC Code) are contained in an appendix.

The 1987 edition of IMOSAR includes earlier amendments.

Although the 'International Conference on Maritime Search and Rescue 1979' (SAR 1979) entered into force on 22 June 1985, ship operators should not assume that there is now a worldwide search and rescue service. The Convention does not stipulate any date by which such a service must become operational. It does, however, describe the way in which an international SAR system should be established. The world's oceans have been divided into 13 SAR areas, some of which by the late 1980s had co-ordinated SAR plans whilst in other areas no plans existed. However, continuing improvements in communication technology will speed the development of co-ordinated SAR plans.

Global maritime distress and safety system (GMDSS)

Technical developments, such as the International Maritime Satellite Organization (INMARSAT), have led to the inception by IMO and other organizations of a global maritime distress and safety system which incorporates technical communication developments.

The basic concept of the system is that both shore and ship rescue authorities will be quickly informed of distress situations and that all relevant units will take part in a carefully co-ordinated rescue operation. IMO has published an A4 size book under the title GMDSS which explains the basic concept of the system, the functions it will perform and how it is expected to be introduced. The book also includes technical descriptions of satellite and terrestrial radio-communication services which will be used in the system, comprising the INMARSAT and COSPAS–SARSAT satellite systems, digital selective calling system, worldwide navigational warning service, including the NAVTEX system, and survival craft transponders for use in SAR operations.

It is anticipated that the GMDSS will be in operation by the early 1990s.

Merchant ship search and rescue manual (MERSAR)

The advantages of modern sophisticated communication systems can be negated if the personnel who are *actually* engaged in a rescue operation are ineffective. In this age of high level shipping technology there is still a very important place for commonsense, practical seamanship. The fourth edition of this manual, published in 1986, states that:

> The purpose of this manual is to provide guidance for those who during emergencies at sea may require assistance from others or who may be able to render such assistance themselves. In particular, it is designed to aid the master of any vessel who might be called upon to conduct SAR operations at sea for persons in distress.

The manual includes sections on co-ordination of SAR operations, actions by ships in distress, action by assisting ships, assistance by SAR craft, planning and conducting the search, conclusion of search, communications, and aircraft casualties at sea.

55

This manual should be carried on the navigation bridge of every vessel. I personally found the MERSAR Manual invaluable when, as Captain of an oil tanker in the South China Sea during a monsoon period, I was required to co-ordinate a SAR operation to rescue the crew of an abandoned log carrier. The operation was brought to a successful conclusion with the aid of MERSAR techniques. All officers should be conversant with the contents of this manual.

Harmonized Survey and Certification System

Three IMO conventions require ships to be surveyed in port or in a repair yard:

> Solas
> Load Lines
> Marpol 73/78

However, the survey intervals of the conventions often do not coincide. Over the operational lifetime of a ship the expense of survey delays can be quite considerable. IMO has acknowledged this problem and consequently has prepared a system whereby the survey and certification requirements of the three conventions will be harmonized, thus reducing delays and costs and in addition making it easier for administrations to ensure that the conventions are being complied with.

It is intended that the system will be in force by 1 February 1992.

SOLAS Ferry Safety Amendments

A series of measures, based on United Kingdom statutory instruments introduced as a result of the *Herald of Free Enterprise* disaster, are contained in a number of amendments to SOLAS 74. The measures were adopted by IMO's Maritime Safety Committee in 1988.

The first series comprise the following changes:

1 The provision of indicators on the bridge to show the proper closure of loading doors and other openings which, if left open, could lead to major flooding.
2 The installation of a surveillance system, such as television monitoring, to detect any leaks through the doors.
3 The patrolling or monitoring of cargo spaces so that the movement of vehicles in bad weather or the presence of unauthorized passengers can be detected.
4 The installation in public spaces and alleyways of supplementary emergency lighting which can operate for at least three hours even if the ship capsizes.

The above measures apply to all ro-ro passenger ships from 22 October 1989, except for a three year period of grace for existing ships with reference to Item 2 and a one year period for Item 4.

Other amendments are operative from 29 April 1990. These amendments affect Chapter II-1 and are summarized below:

(a) An amendment to Regulation 8 will provide more information to the Masters of ro-ro ships with regard to draught, trim and stability after loading and before departure (electronic loading and stability computers may be used for this purpose).

(b) A new regulation, 20–1, requires cargo loading doors to be closed and locked before departure, with an appropriate entry being made in the ship's log.

(c) A new regulation, 22, requires a lightweight survey to be carried out on all passenger ships at intervals not exceeding five years (this will ensure that the stability of ro-ro ships is not adversely affected by changes in weight such as additions to the superstructure).

What could be described as a major change is another amendment to Regulation 8 which is designed to improve the stability of all passenger ships, including ro-ro vessels, in a damaged condition. The amendment represents a major advance in residual stability standards and is intended to ensure that a ship that is damaged to a prescribed extent will remain afloat and stable. The amendment takes into account factors such as wind pressure, passengers crowding on one side and the weight of survival craft being launched and will apply to passenger ships constructed on or after 29 April 1990.

The amendments were adopted under the SOLAS Convention's 'tacit acceptance' procedure which was designed to permit the quick adoption of urgent measures.

E International Publishers of Non-statutory Recommendations

The International Maritime Organization is by far the most important publisher of nautical legislation and codes of practice. However, several industry-based international organizations publish operational guides that are very important and which must not be ignored by ship and shore management. Such guides are particularly valuable as they are written by personnel from the industrial side of the industry rather than from the legislative side and contain much practical, relevant and valuable advice. Three of these organizations are noted below.

The International Chamber of Shipping (ICS)

The Chamber could be said to be the international 'voice' of shipowners as it concentrates on issues that unite shipowners and avoids national or company interests. The Chamber was first formed in 1921 but has increased in strength and stature in the last few decades after recovering from the trauma of the Second World War. The membership is mainly from the long-established maritime nations, and represents about half of the world's merchant tonnage. The ICS has an important role in maintaining and improving the operational standards of shipping and its publications reflect that role. Publications include the *Guide to Helicopter/Ship Operations* and the *Tanker Safety Guide (Chemicals)*.

The Oil Companies International Marine Forum (OCIMF)

The OCIMF was formed in 1970 at a time of increased awareness of the environmental impact of oil pollution. It is an association of oil companies involved with the transport and/or refining of hydrocarbon and associated products. The main objectives of the OCIMF are the promotion of safety and the prevention of pollution, particularly with relevance to the operation of tankers and terminals. The Forum permits the transmission of the opinions and views of oil companies to bodies such as IMO and has an important role in the areas of safety and pollution prevention. The OCIMF publishes guides under its own 'banner', e.g. the *Guide on Marine Terminal Fire Protection and Emergency Evacuation*. However, it is possibly best known for its joint publications, e.g. with the ICS, *Peril at Sea and Salvage: A Guide for Masters*, and with the ICS and the International Association of Ports and Harbours (IAPH), the *International Safety Guide for Oil Tankers and Terminals*.

The Society of International Gas Tanker and Terminal Operators Ltd (SIGTTO)

The Society was founded in 1979 and the members are involved in the operation of gas terminals and/or the operation of gas tankers. One of the main reasons for its existence is therefore the safe and efficient operation of gas tankers and terminals. Although a relatively new body, it has consultative status with IMO and has produced two major reference works, *Liquefied Gas Handling Principles on Ships and in Terminals* and *Recommendations and Guidelines for Linked Ship/Shore Emergency Shut-down of Liquefied Gas Cargo Transfer*.

F   UK Safety Legislation

Statutory Instruments (SIs) and M Notices

The Statutory Instrument is the means by which British Regulations are formulated. Webster's *Third New International Dictionary* defines a statutory instrument as 'a rule, order, or administrative regulation having the force of law'. In 1989 there were over two hundred and forty regulations applicable to British ships.

Merchant Shipping Notices, known as M Notices, are published by the Department of Transport. They are quasi-legal documents in so far as they are not statutory instruments but the recommendations contained in the Notices should be complied with. The M Notices are the method by which the Department of Transport promulgates important information which should be quickly brought to the attention of seafarers and management, or to those associated with the industry. The notices often refer to incidents which have recently occurred and give recommendations to prevent the recurrence of such incidents.

The notices are primarily concerned with safety and new legislation and mariners should have a good working knowledge of the contents of most Notices. The importance of the contents of M Notices should never be underestimated and students in particular should refer constantly to them.

M Notices have the standing of 'authoritative documents', i.e. due consideration must be given to the contents, and offices would be required to account for ignorance of, or failure to comply with, M Notice recommendations at any inquiry concerning a particular incident for which M Notice information was available.

All ships should carry a comprehensive file containing all M Notices which are currently in force. A particular notice appears each year which lists the Notices which are currently in force.

Brief details of some statutory instruments which affect daily operations, or the safe and efficient management of ships, are contained in the following pages.

Life saving

Merchant Shipping (Life-Saving Appliances) Regulations 1980. SI 1980 No. 538

Operative 25 May 1980, these regulations stipulate the life-saving appliances which must be carried by the twenty-one different classes of ship as specified by the regulations. The most common classes of ship are:

I Passenger ships engaged on long international voyages
II Passenger ships engaged on short international voyages
VII Ships other than tankers engaged on voyages any of which are long international voyages
VII(T) Tankers engaged on voyages any of which are long international voyages

The term 'passenger ship' means a ship carrying more than 12 passengers; 'short international voyage' means a voyage in the course of which a ship is not more than 200 miles from a port at which passengers and crew could be landed; a 'long international voyage' is a voyage which is not a short international voyage.

A 'tanker' is a cargo ship which has been constructed or adapted for the carriage in bulk of liquid cargoes of a flammable nature.

The 54 regulations fall into two broad categories:

1 An enumeration of the actual appliances which each class of vessel must have on board.
2 The general requirements with regard to the standards of the appliances.

Twenty-one schedules are appended to the regulations and these schedules give very specific and detailed requirements for particular appliances. For example:

(a) Regulation 11 (2) states that 'every ship to which this regulation applies of 500 tons and over shall carry on each side of the ship one or more lifeboats of sufficient aggregate capacity to accommodate all persons on board'.
(b) Regulation 36 gives details of the equipment and rations that a lifeboat must be provided with.
(c) Schedule 3 gives the construction requirements for lifeboats.

All sea-going officers must have detailed knowledge of the life-saving appliances requirements of SOLAS 83 and also the national requirements for the vessel on which they are actually employed. Officers on British ships must, therefore, carefully study these regulations or the regulations which are mentioned next.

Merchant Shipping (Life-Saving Appliances) Regulations 1986. SI 1986 No. 1066

Operative 1 July, 1986, these regulations apply to 'new ships' (a new ship is a ship the keel of which has been laid on or after 1 July 1986), both to United Kingdom ships wherever they may be and to non-United Kingdom ships while they are within the United Kingdom or the territorial waters thereof.

The main purpose of these regulations is to give effect to the SOLAS 83 Amendments with regard to new British ships. The Merchant Shipping (Life-Saving Appliances Regulations 1980) (Amendment Regulations) 1986, SI 1986 No. 1072 extend to British ships built before 1 July 1986 the SOLAS 83 requirements for:

Manning of survival craft

Provision of training manuals

Requirements for operational readiness, maintenance, inspections and servicing of life-saving appliances

Regulation 28 of the 1986 Regulations states that if a ship proceeds on any voyage without complying with the requirements of the regulations, the owner and master of the ship shall *each* be guilty of an offence and liable on summary conviction to a fine not exceeding £1,000 or, on conviction on indictment, to imprisonment for a term not exceeding two years and a fine.

It is interesting to note that this regulation also states that 'It shall be a good defence to a charge under this regulation to prove that the person charged took all reasonable steps to avoid commission of the offence'. One assumes that the term 'reasonable' would have to be defined by a court of law.

Regulation 29 deals with the 'power to detain' in that any ship which does not comply with the requirements of the regulations is liable to be detained.

Merchant Shipping (Musters and Training) Regulations 1986. SI 1986 No. 1071

Operative 1 July 1986, these regulations give effect to Regulations 8, 18, 50 and 53 of the new SOLAS Chapter III (83 Amendments).

One of the few complaints that I have of Chapter III is the cross-referencing that has to be carried out between different regulations before certain requirements can be fully comprehended. This Statutory Instrument combines the SOLAS musters and training regulations into a well presented and easily understood format, thus making it very useful for examination revision purposes.

The regulations (in general) apply to all United Kingdom ships anywhere in the world and to non-United Kingdom vessels when in British waters.

It should be noted that superficial log entries concerning musters and drills are no longer good enough . Regulation 8 clearly states that:

'(1) The following matters shall be recorded by the master in the official logbook:

(a) upon each occasion on which in accordance with these Regulations, a muster, abandon ship drill, fire drill, drill of other life-saving appliances or on-board training is held:

(i) a record of the date upon which musters, drills and training are held;

      (ii) details of training and type of drill held;

      (iii) a record of the occasions on which lifeboats, rescue boats and davit launched liferafts, as applicable are lowered or launched.

  (b) upon each occasion on which a full muster, drill or training session is not held as required by these Regulations.

      (i) a record of the relevant circumstances;

      (ii) the extent of the muster drill or training session held.

(2) In ships not required to keep an official logbook a record of each matter specified in paragraph (1) shall be made by the master and shall be retained on board for a period of not less than 12 months.'

It should be noted that the Department of Transport is concerned that high safety standards should be maintained and to 'encourage' masters to take such responsibilities seriously these regulations state certain penalties for neglect of duty. If the master of a ship:

(a) does not comply with the muster list and emergency instructions regulations

(b) does not comply with the practice musters and drills regulation

(c) does not comply with the on-board training and instructions regulations
then that master shall be guilty of an offence and liable on summary conviction to a fine not exceeding £1,000 or, on conviction on indictment, to imprisonment for a term not exceeding two years and a fine.

If the master fails to comply with any of the requirements of Regulation 8 he is liable to a fine not exceeding £50.

Any person who fails to carry out the duty assigned to him with regard to muster list duties and any officer who is assigned to ensure that life-saving and fire appliances are maintained in good condition and are ready for immediate use and fails to do so is liable to a fine not exceeding £500.

One would hope that *all* masters would take their safety obligations seriously. Perhaps that is too idealistic a hope as I have been told of several situations in which a master has put pressure on a chief officer to sign official logbook entries with regard to drills which had not actually been carried out because of operational pressures. I would seriously recommend that only correct entries should be made in the logbook.

The term 'summary conviction' applies to offences which are dealt with in a magistrate's court and 'conviction on indictment' applies to offences which are considered by a higher court.

Fire protection

In 1989 there were eight statutory instruments in force which applied to fire protection. The three 'main' instruments are:

(a) Merchant Shipping (Fire Appliances) 'Regulations 1980, SI 1980 No. 544, which apply to ships built on or after the 25 May 1980;
(b) Merchant Shipping (Fire Protection) Regulations 1984, SI 1984 No. 1218, which apply to ships built on or after 1 September 1984;
(c) Merchant Shipping (Fire Protection) (Ships Built Before 25 May 1980) Regulations, SI 1985 No. 1218, which came into operation on 12 August 1985 and which apply to ships built before 25 May 1980.

The regulations are very comprehensive and what might seem to be an excess of regulations is in fact the result of more exacting SOLAS standards and advances in fire protection technology and techniques. It is, therefore, imperative that managers ashore and afloat ensure that all the vessels which they manage comply with the regulations pertaining to individual ships. This is of particular importance when a foreign flag vessel is being registered in the United Kingdom.

The vessel classification system is the same as that specified in life-saving appliances regulations. It should be noted that in the 1984 regulations the emphasis changed somewhat from fire fighting to fire protection. The 1984 statutory instrument contains 147 regulations and 14 schedules which cover three basic 'areas':

Fire prevention
Structural fire protection
Fire appliances

However, some regulations deal with particular items, e.g. Regulation 143 refers to 'Special Requirements for Ships Carrying Dangerous Goods'.

As the regulations are extensively cross-referenced I would recommend that ships' officers study the actual regulations rather than refer to a synopsis.

Management should read carefully Regulations 146 and 147: Regulation 146 states that, if a ship proceeds or attempts to proceed to sea without complying with the regulations, the *owner and master shall each be guilty of an offence* and liable on summary conviction to a fine not exceeding £1,000 or, on conviction on indictment, to imprisonment for a term not exceeding two years and a fine; and Regulation 147 states that *ships are liable to be detained* for non-compliance with the regulations.

M1217 Musters and Drills and On-Board Training and Instruction

This is an important M notice for those officers involved with crew safety training, whether on British or non-British ships. The recommendations and guidance in the notice will greatly assist officers to comply with:

The Merchant Shipping (Musters and Training) Regulations 1986
The Merchant Shipping (Life-Saving Appliances) Regulations 1986

The Merchant Shipping (Life-Saving Appliances) Regulations 1980 (Amendment) Regulations 1986

The Merchant Shipping (Closing of Openings in Hulls and in Watertight Bulkheads) Regulations 1987

SOLAS Chapter III

I have already mentioned the new emphasis on maintenance and training and this notice underlines 'the fundamental importance of training and instruction, in particular on-board training and instruction in the use of a ship's live-saving appliances and in the best methods of survival. Closely associated with such training is the holding of, and taking part in, musters and drills'.

The notice has 15 sections which should be closely studied as, in addition to simply stating the requirements of the regulations, recommendations are given for the practical application of such requirements. For example, Section 7 deals with 'Fire and Other Emergency Drills'. Section 7.1 states that 'A fire or other emergency drill should be held simultaneously with the first stage of the abandon ship drill'. Section 7.2 expands this by advising:

> For the purpose of a fire drill an outbreak of fire should be assumed to have occurred in some part of the ship and a mock attack should be made. The complete co-operation of the personnel of all departments is essential in fire fighting. The type and position of the supposed fire should be varied from time to time and can include:
>
> 1 Cargo fires in holds or other spaces;
> 2 Fires involving oil, gas or chemical cargoes as appropriate;
> 3 Fires in engine or boiler rooms;
> 4 Fires in crew or passenger accommodation;
> 5 Fires in galleys due to burning oil or cooking fats.

Ten more sub-sections continue the advice on planning effective and efficient fire drills.

Section 13 refers to 'Drills in Closing of Doors, Side Scuttles and Other Openings' which are required by the Closing of Openings in Hulls and in Watertight Bulkheads Regulations 1987 and which came into force on 1 November 1987. The notice reminds mariners that those 'Regulations also require inspections, at not more than 7 days, of watertight doors and mechanisms, indicators and warning devices connected with such doors, valves, the closing of which is necessary to make watertight any compartment below the margin line, and valves, the operation of which is necessary for the efficient operation of damage-control cross-connections'.

Masters are reminded of the need to instruct the crew in the operation of such devices and attention is drawn to M notices 1151, 1283 and 1326 which deal with those particular Regulations.

Occupational Health and Safety

Merchant Shipping (Code of Safe Working Practices) Regulations 1980. SI 1980 No. 686

Operative 7 July 1980. Every ship to which these regulations apply which has a total crew (including the master) not exceeding 15 persons must have two copies of the *Code of Safe Working Practices for Merchant Seamen,* one copy to be retained by the master and the other copy kept in a place readily accessible to seamen.

On ships with a crew exceeding 15 persons, one copy of the Code must be kept by the following persons:
(a) The Master
(b) The Chief Officer
(c) The Chief Engineer
(d) The Purser or Catering Officer
(e) The Safety Officer
(f) Safety representatives
(g) Each member of the accident prevention committee who wishes to have a copy

In addition, at least one reference copy should be available for every 25 seamen employed on the ship. At least three notices should state where such reference copies are located.

The master, and any of the persons in the (b) to (f) categories, must make a copy of the Code temporarily available to any seaman on being so requested. Any person who fails without reasonable cause to make available such a copy shall be guilty of an offence and liable on summary conviction to a fine not exceeding £1,000.

The owner of the ship shall ensure that the ship carries sufficient copies of the Code and any person who contravenes this requirement shall be guilty of an offence and liable on summary conviction to a fine not exceeding £1,000.

If the master does not display the notices referred to earlier he is liable to a fine not exceeding £50.

People leaving a ship in order to proceed to a nautical college should bear in mind that 'any person who knowingly removes a copy of the Code . . . from the ship without the consent of the owner or master shall be guilty of an offence and liable on summary conviction to a fine not exceeding £50.'

Merchant Shipping (Safety Officials and Reporting of Accidents and Dangerous Occurrences) Regulations 1982. SI 1982 No. 876

Operative 1 October 1982. Part 1 of the Regulations requires the employer of the crew of every British ship carrying a crew of more than five (with a few

65

exceptions) to appoint a Safety Officer. It also enables the officers and ratings, if they so desire, to elect safety representatives. Once such a representative is elected, the employer is required to appoint a safety committee. The Regulations stipulate that a Safety Officer must 'use his best endeavours to ensure that the provisions of the Code of Safe Working Practices and the employer's occupational health and safety policies are complied with'. Among his many duties he is required to improve the crew's safety awareness, to investigate crew members' safety complaints, to investigate accidents, make recommendations to prevent the recurrence of such accidents, and to carry out inspections. Part 1 also states the powers of a safety representative whereby, on behalf of the crew, he may make representations to the master, the Safety Officer, the employer (through the master), and the safety committee on matters concerning safety. He may also participate in safety investigations or inspections carried out by the Safety Officer, undertake such tasks himself, and request the committee to initiate particular investigations into aspects of safety. The duties of the safety committee include using 'their best endeavours' to improve safety consciousness, to make recommendations and representations to the employer, and to take appropriate action in any health and safety matters which affect the ship and her crew. The duties of the employer and master with regard to access of information and other matters conclude Part 1.

Part 2 of the Regulations provides for the notification of specified accidents and dangerous occurrences to persons employed or carried on board and applies to all United Kingdom ships other than pleasure craft and fishing vessels. The master or, in his absence, the most senior officer must report every accident involving death or serious injury as quickly as possible to the Department of Transport. In addition, if the vessel carries a Safety Officer, the master or the most senior officer shall make a written report of every accident or dangerous occurrence by completing the Department of Transport form, ARF (accident report form, a numerical suffix indicates a revised version of the original form ARF 1, e.g. ARF 2). The master or senior officer, the Safety Officer, and the safety representative must sign the form. Other provisions apply when the vessel is not carrying a Safety Officer.

The Regulations contain a schedule which lists some typical dangerous occurrences which should be reported even if major injury has not occurred, e.g. a derrick collapse, a fall overboard, the parting of a towrope, significant cargo shift etc.

Regulation 12 states that:

'(1) Any person who fails to comply with any of the provisions of these Regulations shall be guilty of an offence and liable on summary conviction to a fine not exceeding £1,000, or on conviction on indictment, to imprisonment for a term not exceeding two years or a fine or both;

(2) In any proceedings for an offence under these Regulations it shall be a defence for the person to show that all reasonable steps had been taken by him to ensure compliance with the Regulations.'

Merchant Shipping (Health and Safety: General Duties) Regulations 1984. SI 1984 No. 408

Operative 24 April 1984, these Regulations set out the general duties of employers and employees with regard to health and safety aboard ship. They apply to United Kingdom ships and to non-British ships in British ports.

An employer is required to ensure, so far as is reasonably practical, the health and safety of his employees and other persons on board ship. This includes:

(a) The provision and maintenance of safe plant, machinery and equipment and systems of work.
(b) Arrangements for the safe use, handling, stowage and carriage of articles and substances.
(c) The provision to employees of necessary health and safety information, instruction, training and supervision.
(d) The maintenance of safe and healthy work places.
(e) The provision and maintenance of a safe and healthy environment.
(f) Collaboration with other employers to facilitate the health and safety of all persons on board ship.

An employer is also required to make a statement of health and safety policy and to bring the statement to the notice of his employees (this shall not apply to an employer who employs less than five employees in aggregate aboard UK ships). No employer is permitted to levy any charge on any employee in respect of these Regulations.

Contravention of the above shall be an offence punishable on summary conviction by a fine not exceeding £1,000.

The employee is required to take reasonable care for the health and safety of all persons on board ship, including himself. He is also required to co-operate with his employer, or any other person, to ensure that any health and safety duties under Merchant Shipping Acts can be carried out. Contravention shall be an offence punishable on summary conviction by a fine not exceeding £50.

Any person charged with contravening the above shall have a defence if he can show 'that he took all reasonable precautions and exercised all due diligence'.

All persons have a duty not to intentionally misuse or recklessly interfere with anything that is provided in the interests of health and safety; contravention shall be an offence punishable on summary conviction by a fine not exceeding £200.

It should be noted that an 'Offences by Body Corporate' regulation is contained in these Regulations which is worth quoting in full:

> 10 (1) Where an offence under any of these Regulations committed by a body corporate is proved to have been committed with the consent or connivance of, or to have been attributable to any neglect on the part of, any director, manager, secretary or other similar officer of the body corporate or a person who was purporting to act in any such capacity, he as well as the body corporate shall be guilty of that offence and shall be liable to be proceeded against and punished accordingly.

Any ship, whether British or not, is liable to be detained in a British port if an inspection by a duly authorized person reveals a failure to comply with the requirements of these Regulations.

These Regulations give effect in part to ILO Convention 147.

Merchant Shipping (Protective Clothing and Equipment) Regulations 1985. SI 1985 No. 1664

Operative 1 May 1986, these Regulations require employers to provide protective clothing and equipment for their employees who are engaged in, or at risk from, hazardous work processes on board ship. The Regulations apply to United Kingdom ships and to non-British ports.

The Regulations make it clear that the provision of clothing and equipment is not sufficient in itself, thus an employer shall ensure that:

1 Every employee engaged in a work process described in M1195 is provided with *suitable* protective clothing and equipment as specified in M1195.
2 Every employee engaged in any other work process involving a particular hazard is similarily provided for.
3 Such equipment is either issued or kept in easily accessible, suitable storage.
4 Such equipment is properly maintained, regularly inspected, checked at intervals of not more than three months, repaired or replaced as necessary and, in the case of breathing apparatus used for work processes mentioned in M1195, inspected and checked before and after use.
5 Employees are instructed in the use of protective clothing and equipment.
6 Instructions for their proper use and maintenance are provided with protective clothing and equipment.

Contravention of the above shall be an offence punishable on summary conviction by a fine not exceeding £1,000. (I would suggest that much of the above could be incorporated into a Planned Maintenance Schedule and also into a Company Crew Training Schedule.)

The Annex to M1195 enumerates the work processes for which protective clothing and equipment must be provided, e.g. 'Any process or activity involving a reasonably foreseeable risk to the head from falling or moving objects' requires 'a general purpose industrial safety helmet' to British Standard specification (M1358 contains amendments to the annex to M1195).

Employees also have duties as each employee:

1 shall ensure that any deficiencies or defects in any items of protective clothing or equipment issued to him for his individual use are reported to a responsible ship's officer; and
2 shall actually wear or use appropriate protective clothing and equipment.

Contravention by an employee of these requirements shall be an offence punishable on summary conviction by a fine not exceeding £100.

No person shall require an employee to start a work process as described by these Regulations unless the appropriate clothing or equipment is provided. Contravention shall be an offence punishable on summary conviction by a fine not exceeding £200.

'All reasonable precautions' and 'all due diligence' is a defence under these Regulations. The Regulations also have a 'body corporate regulation' as described in the previous regulation.

A duly authorized person may inspect and if necessary detain any ship not complying with these Regulations in a British port.

The Regulations give effect in part to ILO Convention 147.

Merchant Shipping (Guarding of Machinery and Safety of Electrical Equipment) Regulations 1988. SI 1988 No. 1636

Operative 1 January 1989, the Regulations apply to United Kingdom ships and to non-British ships in British ports.

The employer and the master shall ensure that:

1 Every dangerous part of the ship's machinery is securely guarded.
2 All guards and similar devices are of substantial construction, properly maintained and in position.
3 There is a means for taking prompt action to stop any machinery and for cutting off the power in the event of an emergency.
4 All ship's electrical equipment and installations are so constructed, installed, operated and maintained that the ship and all persons are protected against electrical hazards.

Contravention of the above by an employer shall be an offence punishable on summary conviction by a fine not exceeding £2,000, or on conviction on indictment by imprisonment for a term not exceeding two years or a fine or both. Contravention by a master shall be an offence punishable only on summary conviction by a fine not exceeding £1,000. 'All reasonable precautions' and exercise of 'all due diligence' is a defence.

A duly authorized person may inspect and, if necessary, detain any ship not complying with these Regulations in a British port.

Regulation 3 (2) deals with procedures when 'the safety of the ship' is paramount and for examination purposes.

M1355 gives some guidance on how the Regulations are to be interpreted. The notice defines machinery as being securely guarded 'if it is protected by a properly installed guard or device which prevents foreseeable contact between a person or anything worn or held by a person and any dangerous part of the machinery'. It is pointed out that it is not the machinery as a whole which needs guarding but only dangerous parts such as gearing, belt drives, reciprocating components and revolving shafts and couplings (the barrels of windlasses, winches and capstans are not normally considered to be dangerous parts nor are mechanical hatch covers). Machines brought on board by dockers or ship repairers in the United Kingdom are subject to the Factories Act or the Shipbuilding and Ship Repairing Regulations. A British Standard (BS 5304) refers to the Safeguarding of Machines.

M1355 also draws attention to the Code, in which further advice is given.

These Regulations give effect in part to ILO Convention 147 and ILO Convention 152, and also allow, in conjunction with other regulations, the repeal of the 1934 Docks Regulations.

Occupational Health and Safety (1988) Regulations in Association with the Docks Regulations (1988)

A major 'package' of regulations came into force on 1 January 1989 which, together with the new Docks Regulations prepared by the Health and Safety Commission, replaces the 1934 Docks Regulations. The package also enables the United Kingdom to ratify the International Labour Organization Convention 152, which is concerned with health and safety in dock work, and give effect to the Merchant Shipping (Minimum Standards) Convention, ILO Convention 147.

The regulations are somewhat of a departure from the traditional style or format of Merchant Shipping statutory instruments. The regulations are brief and are written in general terms but are cross-referenced with particular chapters in the *Code of Safe Working Practices for Merchant Seamen* which contain the 'principles and guidance' to enable the regulations to be fulfilled. I have always advised students to regard the recommendations in the Code as regulations and these new statutory instruments in effect give certain chapters the status of regulations.

The regulations apply to British ships and, in general, to foreign ships in British ports. The 'package' of regulations could be said to be one of the most important sets of British maritime regulations for some years and, as such, should be closely studied by everyone concerned with shipping operations in British ports.

The British regulations are similar to regulations of those countries which ratify ILO conventions.

The Merchant Shipping (Entry into Dangerous Spaces) Regulations 1988. SI 1988 No. 1638

Operative 1 January 1989, the regulations define a dangerous space as:

> any enclosed or confined space in which it is foreseeable that the atmosphere may at some stage contain toxic or flammable gases or vapours, or be deficient in oxygen, to the extent that it may endanger the life or health of any person entering that space

It should be noted that, except when necessary for entry, the master of a ship shall ensure that all entrances to unattended dangerous spaces on a ship are either kept closed or otherwise secured against entry.

The specific duties of personnel are:

(a) the employer shall ensure that procedures for ensuring safe entry and working in dangerous spaces are clearly laid down;

(b) the master shall ensure that such procedures are observed on board the ship; and

(c) no personnel shall enter or remain in a dangerous space except in accordance with the employer's clearly laid down procedures.

The master of (a) any tanker or gas carrier of 500 GRT tons and over, and, (b) any other ship of 1,000 GRT and over shall ensure that drills simulating the rescue of a crew member from a dangerous space are held at intervals not exceeding two months, and that a record of such drills is entered in the official log book. The master must also ensure that any oxygen meter or testing device carried on board is maintained, serviced or calibrated as necessary. A master who contravenes any of the foregoing is liable to be punished on summary conviction by a fine not exceeding £1,000.

In additon to laying down procedures, an employer shall ensure that on each ship where entry into a dangerous space may be necessary an oxygen meter, and such other testing device as is appropriate to the hazard likely to be encountered in any dangerous space, shall be provided. Contravention of these duties by an employer shall be an offence punishable on summary conviction by a fine not exceeding £2,000 or, on conviction on indictment, by imprisonment for a term not exceeding 2 years or a fine or both.

Any person (other than the employer or master) who does not follow correct dangerous space procedures is liable to a fine on summary conviction not exceeding £400.

71

In general, a person who can show that he 'took all reasonable precautions and exercised all due diligence' will have a good defence if charged with any of the above offences.

All ships' officers should carefully note that in recent years all statutory instruments define the term 'master' as including 'any person in charge of the vessel during the absence of the master', thus an officer cannot shift the blame on to a master who is ashore if that officer is neglectful of his duty.

Similarly, employers are being legally required to act in a responsible manner (the term 'employer' means 'the person for the time being employing the master'), e.g. an employer who fails to provide equipment necessary for complying with certain regulations will be guilty of an offence. I suggest that masters ensure that copies of written requests for such equipment are retained, thus if accidents happen because of the absence of required equipment the appropriate people can be held responsible for not supplying the equipment.

Regulations 10 and 11 make it clear that duly authorized persons may inspect *any* ship in British ports and, if necessary, detain the ship until these regulations are complied with.

In order to fulfil the requirements of the regulations personnel 'shall take full account of the principles and guidance contained in the Code'. The relevant information can be found in Chapter 10 of the Code and M notice 1345 contains a copy of Chapter 10. Much of the content of the chapter describes procedures that any competent, professional officer will already be complying with. However, it is a matter of some regret that fatalities over recent years have led to procedures that should be carried out as standard good seamanship and practice being made obligatory by law. The M notice should be carefully studied as adherence to the recommendations will undoubtedly save lives. The chapter defines the basic precautions as:

1 A competent person should make an assessment of the space and a responsible officer to take charge of the operation should be appointed.
2 The potential hazards should be identified.
3 The space should be prepared and secured for entry.
4 The atmosphere of the space should be tested.
5 A 'permit-to-work' system should be used.
6 Procedures before and during the entry should be instituted.

Detailed information on each of these six basic areas is contained in the chapter, together with additional information on topics such as 'drills and rescues', breathing apparatus and resuscitation equipment.

One piece of advice must be strictly followed: 'No one should attempt a rescue without wearing breathing apparatus'.

**The Merchant Shipping (Safe Movement on Board Ship)
Regulations 1988. SI 1988 No. 1641**

Operative 1 January 1989, these regulations place an obligation on both the
master and the employer to ensure that a safe means of access is provided and
maintained to any place on the ship at which a person may be expected to be. In
carrying out the duties arising from these regulations full account must be
taken of the principles and the guidance contained in Chapter 9 of the Code.
Places on the ship at which a person may be expected to be, include accommodation areas as well as normal place of work. 'Persons' in the context of these
regulations include passengers, dock-workers, and other visitors to the ship
on business but exclude persons who have no right to be on the ship.

The employer and master shall ensure:

1 That safe means of access is provided and maintained to any place on the
   ship to which a person may be expected to go.
2 That all deck surfaces used for transit about the ship, and all passageways,
   walkways and stairs, are properly maintained and kept free from materials or
   substances liable to cause a person to slip or fall.
3 That those areas of the ship being used for the loading or unloading of cargo
   or for other work processes or for transit are adequately and appropriately
   illuminated.
4 That any permanent safety signs used on board the ship for the purpose of
   giving health or safety information or instruction comply with British Standard 5378 Part 1 or with any equivalent standard.
5 That any opening, open hatchway or dangerous edge into, through or over
   which a person may fall is fitted with secure guards or fencing of adequate
   design and construction, which shall be kept in a good state of repair.
6 That all ship's ladders are of good construction and sound material, of
   adequate strength for the purpose for which they are used, free from patent
   defect and properly maintained.
7 That no ship's powered vehicle or powered mobile lifting appliance is driven
   in the course of a work process except by a competent person who is authorized to do so.
8 That danger from use or movement of all such vehicles and mobile lifting
   appliances is so far as is reasonably practicable prevented.
9 That all ship's vehicles and mobile lifting appliances are properly maintained.

Contravention of any of the above by an employer shall be an offence
punishable on summary conviction by a fine not exceeding £2,000 or on
indictment by imprisonment for a term not exceeding 2 years or a fine, or both.
Contravention of any of the above by a master shall be an offence punishable
only on summary conviction by a fine not exceeding £1,000.

In addition, the owner shall ensure that in a new ship (a ship the keel of which was laid on or after 1 January 1989) ladders providing access to the hold comply with the requirement specified in paragraph 6.4 of Chapter 9 of the Code. Contravention shall be an offence punishable as above.

Once again, a person who has taken all reasonable precautions and has exercised all due diligence will have a good defence if charged with contravention of any of the above.

M notice 1344 contains Chapter 9 of the Code. All of the chapter and the M notice must be carefully studied. For example, Section 5 deals with 'Guarding of Openings':

5.2 Any hatchway open for the purposes of handling cargo or stores through which a person may fall should be closed as soon as those operations cease, except during short interruptions of work, including meal breaks, or where closure cannot be effected without prejudice to safety or mechanical efficiency because of the heel or trim of the ship.

5.3 The guardrails or fencing should be free from sharp edges and should be properly maintained. Where necessary, locking devices, and suitable stops or toe-boards should be provided. Each course of rails should be kept substantially horizontal and taut throughout their length.

5.4 Guardrails or fencing should consist of an upper rail at a height of 1 metre and an intermediate rail at a height of 0.5 m. The rails may, where necessary, consist of taut wire or taut chain. Where existing fencing to a height of at least 920 mm has been provided this need not be replaced while it remains secure and adequate.

The M notice also deals with topics such as drainage and watertight doors and gives general advice to seafarers.

Ships are liable to inspection and may be detained until the 'health and safety of all employees and other persons aboard ship is secured'.

The Merchant Shipping (Means of Access) Regulations 1988. SI 1988 No. 1637

Operative 1 January 1989.

Duties of the employer and master

1 Ensure that there is a safe means of access between the ship and any quay, pontoon or similar structure or another ship alongside which the ship is secured.
2 Ensure that any equipment necessary to provide a safe means of access is placed in position promptly after the ship has been secured and remains in position while the ship is secured.
3 Ensure that access equipment which is in use is properly rigged, secured, deployed and is safe to use and is so adjusted from time to time as to maintain safety of access.

4 Ensure that access equipment and immediate approaches thereto are adequately illuminated.

5 Ensure that any equipment used for means of access and any safety net is of good construction, of sound material, of adequate strength for the purposes for which it is used, free from patent defect and properly maintained.

6 When access is necessary between ship and shore, and the ship is not secured alongside, the employer and master shall ensure that such access is provided in a safe manner.

7 Ensure that a life-buoy with a self-activating light and also a separate safety line attached to a quoit or some similar device is provided ready for use at the point of access aboard the ship.

8 Ensure that an adequate number of safety nets is carried on the ship or is otherwise readily available.

9 Ensure that a portable ladder is used for the purpose of access to the ship only where no safer means of access is reasonably practical.

10 Ensure that a rope ladder is used only for the purpose of access between a ship with high freeboard and a ship with low freeboard or between a ship and a boat if no safer means of access is reasonably practicable.

The employer and the master shall take full account of the principles and guidance contained in Chapter 8 of the Code.

Duties of the employer alone

1 In every ship of 30 metres or more registered length ensure that there is carried on the ship a gangway which is appropriate to the deck layout, size, shape and maximum freeboard of the ship and which complies with the specifications set out in British Standards Institution BSMA 78: 1978.

2 In every ship of 120 metres of more in registered length ensure that there is carried on the ship an accommodation ladder which is appropriate to the deck layout, size, shape and maximum freeboard of the ship and which complies with the specifications set out in BSMA 89: 1980.

Duties of the master alone

The master shall ensure that when access equipment is in use and there is a risk of a person falling from that access equipment or from the ship or from the quayside immediately adjacent to the access equipment, a safety net is mounted in order to minimize the risk of injury.

Contravention of a master's duties shall be an offence punishable only on summary conviction by a fine not exceeding £1,000.

Contravention of an employer's duties shall be an offence punishable on summary conviction by a fine not exceeding £2,000 or on conviction on indictment by imprisonment for a term not exceeding two years or a fine, or both.

When access equipment is provided in accordance with these regulations any person boarding or leaving the ship shall use that equipment except in emergencies; failure to do so shall be an offence punishable on summary conviction by a fine not exceeding £400.

A person who took all reasonable precautions and exercised all due diligence would have a good defence if charged under these regulations.

Any ship which does not comply with the requirements of these regulations is liable to be detained until the health and safety of all employees and other persons aboard ship is secured.

M notice 1343 contains Chapter 8 of the Code.

Gangways

BSMA 78: 1978 contains the specification for 'Aluminium shore gangways' and a few of the provisions are set out below. Each gangway is tested when fully assembled.

1 Decking shall comprise either continuous flat-topped longitudinal section or individual flat-plate surfaces, which may have a non-slip coating.
2 Footsteps of 50 × 50 mm hardwood shall be secured to the top of the decking at regular centre-to-centre intervals of not less than 400 mm. Footsteps may also be constructed from extruded aluminium sections of particular specifications.
3 Stanchions are of carbon steel or aluminium to a height of 1,000 mm, fitted at regular intervals (maximum interval 1,500 mm) and vertical to the horizontal gangway. The stanchions shall be one of three types:
   (a) fixed
   (b) hinged, but with securement to prevent inadvertent collapse
   (c) portable, with securement to prevent accidental displacement from the socket
4 Hand and intermediate guides shall be two in number at a height of 1,000 mm and 500 mm respectively.
   They may be of the following types:
   (a) continuous and adequately tensioned sisal, manila or polypropylene or plastic covered wire rope of 16 mm diameter
   (b) galvanized steel chain provided with adequate means of tensioning
   (c) continuous rigid guides of particular specifications
5 Toeboards 150 mm high shall be fitted on each side.
6 A roller or wheels of 110 mm minimum diameter to be fitted on one end.
7 Suitable attachments must be provided for the securing ropes.
8 Four lifting lugs must be provided.
9 Width to be 600 mm.
10 The construction shall allow for an angle of use up to 30° from the horizontal. However, a purchaser can specify a maximum angle to the manufacturer.

Each gangway shall be permanently marked with the manufacturer's name, model number, maximum designed angle of use, and the maximum safe loading by number of persons and by total weight.

Accommodation ladders

Accommodation ladders must conform to the specifications set out in Standard BSMA 89: 1980 of which a few provisions are noted below. The materials used are generally steel or aluminium of suitable standards. Each ladder must support a test load.

1 Ladders may be of single-flight or multi-flight construction and are of two basic types:
   (a) The revolving-platform ladder is capable of being varied in direction and inclination between the ship and the lower access level. The ladder may be supported by steel wire ropes or chains or by rollers fixed to the bottom of the ladder.
   (b) The fixed-platform ladder is hinged from a fixed anchorage and is capable of being varied in inclination only. Support is provided by steel wire ropes or chains.
2 The distance between the steps shall be 300 mm (measured tangential to the step noses).
3 The top handrail should be at a height of 1,000 mm and an intermediate rail should be provided at mid-height.
4 The width of all ladders shall be 600 mm.
5 Ladders shall be capable of being operated safely in a horizontal position and shall operate at an angle of 55° with the steps horizontal.
6 All the support points, such as pivots, and suspension points, such as lugs, shall be of adequate strength to support the weight of the ladder and the design load.

Each ladder shall be permanently marked by the manufacturer's name, model number, the maximum designed angle of use, and the maximum safe loading by numbers of persons and total weight.

Chapter 8 states that 'Gangways should not be used at an angle of inclination greater than 30° from the horizontal and accommodation ladders should not be used at an angle greater than 55° from below the horizontal, unless specifically designed for greater angles.' M 1343 should be carefully studied, but a few points are of particular note.

Bulwark ladders. When the inboard end of the gangway or accommodation ladder rests on or is flush with the top of a bulwark, a bulwark ladder, which complies with the specifications set out in Shipbuilding Industry Standard No SIS7, should be provided. Adequate fittings shall be provided to enable the bulwark ladder to be properly and safely secured.

Any gap between the bulwark ladder and the gangway or accommodation ladder should be adequately fenced to a height of at least 1 metre.

Portable and rope ladders. When it is necessary to use a portable ladder for access, as described earlier, it should be used at an angle of between 60° and 75° from the horizontal. The ladder should extend at least 1 metre above the upper landing place unless there are other suitable handholds. It should be properly secured against slipping or shifting sideways or falling and be so placed as to afford a clearance of at least 150 mm behind the rungs. When a portable ladder is rested against a bulwark or rails, a bulwark ladder as detailed above should be used.

When it is necessary to use a rope ladder, as described earlier, it should never be secured to rails or to any other means of support unless the rails or support are so constructed and fixed as to take the weight of a man and a ladder with an ample margin of safety. A rope ladder should be left in such a way that it either hangs fully extended from a securing point or is pulled up completely. It should not be left so that any slack will suddenly pay out when the ladder is used. A rope ladder shall be of adequate width and length and so constructed that it can be efficiently secured to the ship. The steps shall provide a slip-resistant foothold of not less than 400 mm × 115 mm and shall be so secured that they are firmly held against twist, turnover or tilt. The steps shall be equally spaced at invervals of 310 mm ($\pm$5 mm). Ladders of more than 1.5 metres in length shall be fitted with spreaders not less than 1.8 m long. The lowest spreader shall be on the fifth step from the bottom and the interval between spreaders shall not exceed nine steps.

Safety nets. Where there is a risk of a person falling from the access equipment or from the quayside or ship's deck adjacent to the access equipment, a safety net shall be mounted where reasonably practical. The aim of safety nets is to minimize the risk of injury arising from falling between the ship and quay or falling onto the quay or deck and as far as reasonably practical the whole length of the means of access should be covered. Safety nets should be securely rigged, with use being made of attachment points on the quayside where appropriate.

General points
1 In normal circumstances, the whole means of access and the immediate approaches should be effectively illuminated from the ship or the shore to at least the level of 20 lux, as measured at a height of 1 metre, above the surface of the means of access or its immediate approaches (a minimum level of 30 lux should be considered when conditions warrant it, e.g. in the presence of coal dust).
2 When the access equipment is provided from the shore it is *still the responsibility of the master* to ensure as far as is reasonably practical that proper standards are maintained.

3 All access equipment should be inspected by a competent person at appropriate intervals and defects should be reported and made good. Rigging should be taut at all times.

4 No access equipment should be painted or treated to conceal any cracks or defects.

5 Aluminium equipment should be examined for corrosion in accordance with the instructions in Annex 1 of the Code.

6 Gangways and other access equipment should not be rigged on ships' rails unless the rail has been reinforced for that purpose.

7 The means of access should be sited clear of the cargo working area and so placed that no suspended load passes over it, when this is not practicable access should be supervised at all times.

8 The means of access and its immediate approaches should be kept free from obstruction and kept clear of any substances likely to cause a person to slip or fall.

The above is a synopsis of Chapter 8 and M1343 *must* be closely studied.

Copies of the appropriate standards can be obtained from the British Standards Institution, Linford Wood, Milton Keynes, MK14 6LE, United Kingdom.

The Merchant Shipping (Hatches and Lifting Plant)—Regulations 1988. SI 1988 No. 1639

Operative 1 January 1989. It should be noted that within the context of these regulations a 'competent person' means a person over the age of 18 possessing the knowledge and experience required for the performance of thorough examinations and tests of ships' lifting plant. The term 'thorough examination' is defined in Chapter 17 of the Code as 'a detailed examination by a competent person, supplemented by such dismantling as the competent person considers necessary, and access to or removal of hidden parts also at the discretion of the competent person in order to arrive at a reliable conclusion as to the safety of the plant examined'.

Lifting plant
Duties of the employer and master

1 Ensure that any lifting plant is of good design, of sound construction and material, of adequate strength for the purpose for which it is to be used, free from patent defect, properly installed or assembled and properly maintained.

2 Ensure that lifting plant is not used other than in a safe and proper manner.

3 Ensure that except for the purpose of carrying out a test the lifting plant is not loaded in excess of its safe working load (refer to page 131 for test details).

4 Ensure that no lifting plant is used
   (a) after manufacture or installation, or
   (b) after any repair or modification which is likely to alter the safe working
       load or affect the lifting plant's strength or stability,
   without first being suitably tested by a competent person (after 1 January
   1993 ensure that a lifting appliance is not used unless it has been suitably
   tested by a competent person within the preceding five years).
5 Ensure that any lifting plant is not used unless it has been thoroughly
   examined by a competent person
   (a) at least once in the preceding 12 month period, and
   (b) following a test as above.
6 Ensure that a certificate or report in a form approved by the Secretary of
   State is obtained within 28 days following any test or examination and is
   kept in a safe place on board ship for a period of at least 2 years from receipt
   of the certificate or report of the *next following* test or examination.
7 Ensure that each lifting appliance is clearly and legibly marked with its safe
   working load and a means of identification.
8 Ensure that any crane that is carried on the ship and whose safe working
   load (SWL) varies with its operating radius is fitted with an accurate indicator, clearly visible to the driver, showing the radius of the load lifting
   attachment at any time and the SWL corresponding to that radius.
9 Ensure that each item of lifting gear is clearly and legibly marked with its
   SWL and a means of identification.
10 Ensure that each item of lifting gear which weighs a significant proportion
   of the SWL of any lifting appliance with which it is intended to be used is,
   an addition to its SWL, clearly marked with its weight.

Every employer and master shall take full account of the principles and
guidance in Chapter 17 of the Code.

In addition to the above the master shall ensure that any pallet or similar
piece of equipment for supporting loads or lifting attachment which forms an
integral part of the load of a one-trip sling or pre-slung cargo sling is not
used on a ship unless it is of good construction, of adequate strength for the
purpose for which it is used and free from patent defect (the term 'one-trip
sling' means a sling which has not previously been used for lifting any other
load and is fitted to the load at the commencement of the journey and
intended to be disposed of at the destination of that journey).

Contravention of the above by an employer shall be an offence punishable on
summary conviction by a fine not exceeding £2,000 or on conviction on
indictment by imprisonment for a term not exceeding 2 years or a fine, or both.

Contravention of the above by a master shall be an offence punishable only
on summary conviction by a fine not exceeding £1,000.

Duties of other personnel

No person shall operate any lifting plant unless he is trained and competent to do and has been authorized by a responsible ship's officer. Any person concerned with carrying out any obligations with regard to lifting plant shall take full account of the principles and guidance in Chapter 17 of the Code. Contravention of this paragraph shall be an offence punishable only on summary conviction by a fine not exceeding £400.

M1347 contains a copy of Chapter 17 of the Code. All personnel concerned with ships' lifting plant should carefully study the chapter but it is *essential* that all ships' officers study the 115 procedures detailed in the chapter. Some of the procedures could be classed as 'innovative' but much of the chapter is concerned with 'established practice' (my terminology). An example of the former concerns the regulation which states that 'No person shall operate any ship's lifting plant unless he is trained and competent to do so and has been authorized by a responsible ship's officer' (Regulation 6(5)).

Chapter 17 contains the following recommendations:

1 Training should consist of theoretical instruction to the extent necessary to enable the trainee to appreciate the factors affecting the safe operation of the lifting plant, ship's ramp or retractable car-deck and of practical work with the appropriate plant etc. under supervision.
2 For a person under 18 years of age undergoing training, the degree of direct supervision required should be related to the trainee's experience, perceived competence and the nature of the appliance etc. on which he is being trained. Any work he carries out should be part of his training.
3 After training each person should undergo a test, and if he passes should be given a certificate specifying the type of appliance on which the test was carried out.
4 Where persons have been regularly authorized to operate a class of lifting plant for a period of at least 2 years before the above regulations became operative they may be considered competent for the award of a certificate, provided there is no reason to believe otherwise.
5 Employers should keep records of training and testing undertaken and should ensure the routine monitoring of the competence of those operating lifting appliances.

The operational guidance for the conventional type of ship's derrick could be said to be 'established practice'. The guidelines include the following points:

1 Runners should be fitted to all derricks so that when the runner is slack the bight is not a hazard to persons walking along the deck.
2 Before a derrick is raised or lowered all persons in the vicinity should be warned not to stand in wire bights.

3 All necessary wires should be flaked out.

4 When a single-span derrick is being raised, lowered or adjusted, the hauling part of the topping lift or bull-wire (i.e. winch end whip) should be adequately secured to the drum end.

5 The winch-driver should raise or lower the derrick at a speed consistent with the safe handling of the guys.

6 Before a derrick is raised, lowered or adjusted with a topping lift purchase, the hauling part of the span should be flaked out for its entire length in a safe manner. A seaman should back up to assist the man controlling the wire on the drum by keeping the wire clear of turns and in making fast to the bitts or cleats.

7 To ensure that the derrick is in its final position, the topping lift purchase should be secured to bitts or cleats by first putting on three complete turns followed by four crossing turns and finally securing the whole with a lashing to prevent the turns jumping off due to the wire's natural springiness.

8 When a derrick is lowered on a topping lift purchase, a seaman should be detailed for lifting and holding the pawl bar ready to release it should the need arise.

Particular attention should be paid to Chapter 17, 2.43: 'A load greater than the safe working load may be applied to lifting plant only for the purpose of a test'.

Personnel should be aware that these regulations are not solely concerned with cargo handling appliances and are much more widely applicable than the 1934 Docks Regulations which they replace. The terms 'lifting appliance', 'lifting gear' and 'lifting plant' must be clearly understood. Regulation 2 gives the following definitions:

> Lifting appliance means any ship's stationary or mobile appliance (and every part thereof including attachments used for anchoring, fixing or supporting that appliance but not including vehicle coupling arrangements) which is used on a ship for the purpose of suspending, raising or lowering loads or moving them from one position to another whilst suspended and includes ship's lift trunks and similar vehicles; it does not include—
>> (a) pipes, or gangways; or
>> (b) screw, belt, bucket or other conveyors;
> used for the continuous movement of cargo or people but does include the lifting appliances used to suspend, raise, lower or move any of these items;
>> (c) survival craft or rescue boat launching and recovery appliances or arrangements; or
>> (d) pilot hoists.
> Lifting gear means any gear by means of which a load can be attached to a lifting appliance and which does not form an integral part of that appliance

or load but does not include pallets, one-trip slings and pre-slung cargo slings and freight containers.

Lifting plant includes any lifting appliance or lifting gear.

Hatches
Duties of the employer and master

Ensure that any hatch covering used on a ship is of sound construction and material, of adequate strength for the purpose for which it is used, free from patent defect and properly maintained.

Duties of the master

1 Ensure that a hatch covering is not used unless it can be removed and replaced, whether manually or with mechanical power, without endangering any person.
2 Ensure that information showing the correct replacement position is clearly marked, except in so far as hatch coverings are interchangeable or incapable of being incorrectly replaced.
3 Ensure that a hatch is not used unless the hatch covering has been completely removed, or if not completely removed, is properly secured.

Every employer, master and person carrying out the obligations with regard to hatches shall take full account of the principles and guidance in Chapter 18 of the Code.

Except in the case of an emergency endangering health or safety, no person shall operate a hatch covering which is power-operated or a ship's ramp or a retractable car-deck unless authorized to do so by a responsible ship's officer.

Contravention of these regulations by an employer shall be an offence punishable on summary conviction by a fine not exceeding £2,000 or on conviction on indictment by imprisonment for a term not exceeding 2 years or a fine, or both.

Contravention of these regulations by a master shall be an offence punishable only on summary conviction by a fine not exceeding £1,000.

Any other person who fails to take full account of the principles and guidance in Chapter 18 of the Code or who operates a powered hatch covering or a ship's ramp or retractable car-deck without authorization shall be liable to a fine not exceeding £400 on summary conviction.

M1346 contains Chapter 18 of the Code and once again this M notice should be closely studied by all personnel involved in any way with ships' hatches (this includes Port and Harbour Authority personnel and stevedores).

In general, a person who can show that he 'took all reasonable precautions and exercised all due diligence' will have a good defence if charged with any of the offences in these regulations.

Any ship, when inspected by a duly authorized person, which does not comply with the requirements of these regulations is liable to be detained until the health and safety of all employees and other persons aboard ship is secured.

Merchant Shipping (Safety at Work Regulations) (Non-UK Ships) Regulations 1988

Operative 1 January 1989. The British Department of Transport has made it very clear by issuing this Regulation that British Safety at Work Regulations must be complied with by non-UK ships when in British ports. Criminal proceedings can now be taken by the Department of Transport against employers and masters associated with foreign ships which breach the following 1988 merchant shipping regulations:

> Guarding of Machinery and Safety of Electrical Equipment
> Means of Access
> Entry into Dangerous Spaces
> Hatches and Lifting Plant
> Safe Movement on Board Ship

Each of these individual regulations contains a regulation which permits non-UK ships in British ports to be inspected by Department of Transport surveyors and if the standards are not good enough a report can be sent to the government of the country in which the ship is registered. The ships can also be detained if the conditions warrant it.

However, in addition to the above this Regulation now makes it possible to directly punish an employer and/or master. An employer is now liable on summary conviction to a fine not exceeding £2,000 or on conviction on indictment to imprisonment for a term not exceeding two years or a fine or both.

In practice it may be difficult for anyone to actually 'lay hands' on an employer and it is more likely that the master will suffer for contravention of safety regulations. The master of a non-UK ship is liable on summary conviction to a fine not exceeding £1,000. I would advise masters of ships trading to British ports to study carefully the occupational health and safety regulations.

The Docks Regulations 1988. SI 1988 No. 1655

Operative 1 January 1989. These Regulations were greatly welcomed as the 1934 Docks Regulations which they replaced had been widely regarded as 'out of date' for a long time. The new Regulations are based on modern dock procedures and the ILO Convention 152 on Health and Safety in Dock Work (1979) and are complementary to merchant shipping regulations. In the past there has been some confusion as to which regulations apply to 'the ship' and which to 'the shore', but the '1988 package' of regulations clearly lays down which regulations apply to ships and which to shoreside operations when dealing with interface between ship and shore. These Regulations make it clear that they are applicable *only to dock operations* within Great Britain (Northern Ireland has parallel regulations and for practical purposes one can interchange the terms GB and UK within the context of these Regulations). The terms 'dock operations' and 'dock premises' should be carefully studied but in general:

Dock operations include the loading or unloading of goods on or from a ship and movement of passengers on or from a ship at dock premises and any incidental activities such as the movement of goods, passengers or vehicles, the mooring of a ship, the storing, checking, sorting, inspecting, weighing or handling of goods, fuelling and provisioning of a ship, etc.

Dock premises are 'any dock, wharf, quay, jetty or other place at which ships load or unload goods or embark or disembark passengers, together with neighbouring land or water which is used or occupied, or intended to be used or occupied, for those or incidental activities, and any part of a ship when used for those or incidental activities'.

The scope of the Regulations is thus very wide and all personnel involved with ship/shore interface operations should carefully study them. In general the Regulations deal with the health, safety and welfare requirements of dock operations but the basic areas covered could be defined as:

Planning and control of dock operations
Lighting
General access, both by land and water
Rescue, life-saving and fire-fighting equipment and means of escape
Hatches, ramps and car-decks
Drivers of vehicles and lifting appliances operators
Use of vehicles
Use of lifting plant
Testing, examination, marking and documentation of lifting plant
Confined spaces
Welfare amenities
Protective clothing

The Regulations have been made under the Health and Safety at Work Act 1974 and the Health and Safety Executive is responsible for the enforcement of statutory provisions in relation to any activity carried on in dock premises.

The Docks Regulations are now of reduced importance to mariners as many of the regulations which were previously in the 1934 Docks Regulations are now contained in Merchant Shipping regulations. The Department of Transport and the Health and Safety Commission have thus reduced the possibility of confusion between which is 'ship' responsibility and which is 'shore' responsibility. A knowledge of the Statutory Instrument in itself will possibly contain enough information for ships' staff but it will certainly be insufficient for shore staff.

The HSC has produced an excellent publication entitled *Safety in Docks, Docks Regulations 1988 and Guidance—Approved Code of Practice* which can be obtained from the HMSO Publications Centre, PO Box 276, London SW8 5DT at a modest cost. This is the shore equivalent to the '1988 Merchant Shipping package' and is actually another 'package' as it contains the Regulations, an approved code of practice and guidance notes; however, in this case the package is contained in one publication. All managers who are involved in any aspect of dock operations should carefully study this publication.

'Herald of Free Enterprise' Legislation

The 7,951 GRT ro-ro car passenger ferry *Herald of Free Enterprise* capsized on 6 March 1987 whilst leaving Zeebrugge, with the loss of 150 passengers and 38 members of her crew.

Since the mid-1970s many people had expressed concern over various aspects of the design and operation of ro-ro vessels. The design problems included doubts about stability after hull penetration, the integrity of hulls, watertight integrity of opening arrangements, and free surface effect on large undivided car decks if water got inside the hull. Operational problems included the lack of stability checks when loaded, difficulty of ascertaining the actual number of passengers on board, lashing of vehicles, and the ascertaining of the weight of vehicles and loads. In my opinion both the government and owners were slow to face up to the problems of these vessels, and as a frequent passenger on such ships I believed that 'there was a disaster just waiting to happen'. Unfortunately for 188 people and their relations and friends the disaster did indeed happen.

The *Herald* was lost because one of the basic tenets of seamanship was ignored – the vessel left the berth with her watetight integrity impaired. The ship sailed with her inner and outer bow doors open, water from the bow wave entered a car deck and due to free surface effect caused the ship to lurch quickly to port to an angle of 30^0, from which position she capsized.

All senior ship's officers and shore management should study the *Report of the Court of Inquiry into the Loss of the Herald of Free Enterprise* (HMSO 1987). The report contained what could be construed as a very strong observation in that 'from top to botton the body corporate was infected with the disease of sloppiness'. My sea career was with companies which gave careful consideration to constructive suggestions from sea staff, but my teaching experience indicates that this is not always the case. I have always advised senior officers to keep written copies of all 'safety' requests and comments sent to shore management. The Court of Inquiry was told of specific requests from masters that were ignored by shore management. These included:

(a) lights to be fitted on the bridge to indicate whether bow and stern doors were open or closed;

(b) means to ascertain whether an excess number of passengers was being carried; and

(c) ships to be provided with instruments for reading draughts.

The report shows that shore managers who ignore the advice of their skilled, experienced, professional sea staff are acting improperly. Traditionally the ship's master has been largely held to account when things have gone wrong, but a study of recent legislation indicates that an increasing onus is being placed on the duties of the 'body corporate' and of the owner. It is possible that we may see employers being imprisoned for failure to comply with statutory regulations.

The British Government is to be commended for quickly making legislation to reduce the operational problems which were highlighted by the Court of Inquiry Report. However, one must wonder why often a disaster has to occur before obvious safety deficiencies are corrected (this is not just a marine transport problem, the Manchester Airport disaster in 1985 and the London Underground disaster of 1987 also highlighted safety deficiencies which had been ignored).

Merchant Shipping (Passenger Boarding Cards) Regulations 1988. SI 1988 No. 191

Operative 29 February 1988, these Regulations apply to United Kingdom passenger ships when operating as passenger ships of Classes II and IIA (basically ro-ro ferries on short international voyages). They require the operation of a passenger boarding card system with documentation to be retained for future inspection by duly authorized persons. It is an offence for a ship to leave her berth before the total number of passengers on board has been determined and the master informed.

Duties of the owner

1 The owner shall ensure that there is a system of *individual* passenger boarding cards as described in M1312 (or subsequent M notices).

2 No passenger ship shall leave her berth before the total number of passengers on board has been determined by means of the boarding card system and the master informed (breach of 1 and 2 can lead to a fine not exceeding £2,000 or on conviction on indictment to imprisonment for a term not exceeding two years or a fine or both).

3 No passenger shall be permitted to board unless issued with a boarding card.

4 Arrangements shall be made to determine the number of passengers remaining on board from the previous voyage, such arrangements to be descibed in written instructions.

5 Relevant documents to be retained either on board or ashore and to be available for inspection after being sealed (breach of 3, 4 and 5 can lead to a fine not exceeding £1,000).
   Note that the term 'owner' is used and not the 'employer'.

Duties of the master

1 To ensure that the ship shall not leave her berth before the total number of passengers on board has been determined by means of the boarding card system and he has been informed of that number (breach of this requirement can lead to a fine not exceeding £2,000 or on conviction on indictment to imprisonment for a term not exceeding two years or a fine or both).
2 No passenger shall be permitted to board unless issued with a boarding card.
3 Arrangements shall be made to determine the number of passengers remaining on board from the previous voyage which are described in written instructions (breach of 2 and 3 can lead to a fine not exceeding £1,000).

Any person who knowingly or recklessly makes any false statement in connection with the system of individual passenger boarding cards liable or intended to lead to error in the determination of the total number of passengers or falsifies records or documents or the stealing of such records or documents shall be guilty of an offence and liable on summary conviction to a fine not exceeding £2,000 or on conviction on indictment to imprisonment for a term not exceeding two years or a fine or both.

'All reasonable steps to avoid commission' of any offence shall be a good defence. These Regulations contain an 'Offence by Bodies Corporate' regulation.

M1312 should be closely studied as it contains much practical guidance for the actual operation of the passenger boarding card system. The notice states that in framing the Regulations the Department of Transport had two objectives in mind:

(a) The system of boarding cards to be adopted should be simple and clear, so that it will remain effective in the years to come when the recent casualty, which promted the introduction of the requirements, is no longer uppermost in the operators' minds
(b) The proper functioning of the system should be capable of being easily checked and, if necessary, enforced by the Department's surveyors without the necessity of arranging a full passenger count as passengers leave the ship

It should be noted that the Merchant Shipping (Passengers Boarding Cards) (Application to non-UK Ships) Regulations 1988 SI 1988 No.641, operative 1 April 1988, extended the above Regulations to foreign flag vessels operating to and from British ports.

Merchant Shipping (Closing of Openings in Enclosed Superstructures and in Bulkheads above the Bulkhead Deck) Regulations 1988. SI 1988 No. 317

Operative 9 March 1988. It should be noted that within the context of these Regulations a voyage commences when a ship leaves her berth or anchorage at a port.

These Regulations must be carefully studied as only some of the more pertinent ones are mentioned below.

Regulation 2 states that 'the following loading doors:

(a) gangway and cargo loading doors fitted in the shell or boundaries of enclosed superstructures,
(b) bow visors so fitted,
(c) weathertight ramps so fitted and used instead of doors for closing openings for cargo or vehicle loading,
(d) cargo loading doors in the collision bulkhead, shall be closed and locked before the ship leaves its berth and shall be kept closed and locked until the ship has been secured at its next berth'.

Regulation 3 deals with the supervision and report of closure of doors and clearly states that an officer who has been appointed by the master shall verify that every loading door has been closed and locked and report this to the bridge *before the ship proceeds on a voyage.*

Regulation 4 deals with the closure of watertight or weathertight doors above the margin line and the provisions are similar to those for loading doors. ('Watertight' generally applies when there is a posssibility of water accumulating at either side; 'weathertight', when water may accumulate at one side only).

In general, entries shall be made in the official log book (OLB) to record the times of opening and closing the above doors.

The owner is required to provide what is known as a 'Berth List' which lists the individual berths the ship may use, and the ship shall only use those berths except in a case of emergency.

Regulation 8 requires the owner to supply written instructions, approved by the Secretary of State, concerning the doors to which the Regulations apply and none of the doors shall be opened or closed except in compliance with the written instructions. The instructions shall be kept on the ship at all times in the custody of the master, and shall include the following information:

(a) The circumstances in which the doors to which these Regulations apply may be opened and are required to be closed
(b) A list of the small doors included in this Regulation
(c) The requirements of verifying and reporting the closure of doors
(d) Procedures for opening doors in an emergency

(e) The entries required to be made in the OLB
(f) A reference to the Berth List (which may have additional necessary information with it) and a clear statement that loading and discharging may be done only at the berths listed
(g) Warning of the penalties for not complying with these Regulations

Anyone guilty of an offence shall be punished 'on summary conviction by a fine not exceeding £2,000 or, on conviction on indictment, to imprisonment for a term not exceeding two years *and* a fine'. (Note that it is not 'or' a fine but 'and' a fine.) 'All reasonable steps' is a good defence. These Regulations contain a 'corporate body' offence, regulation (11).

It should be noted that Regulation 2 permits a vessel to move a short distance from the cargo loading/discharging position (not more than one ship's length) before closing a bow visor or a weathertight ramp if that operation cannot be done at the berth. The same provision applies with regard to opening the same.

These Regulations apply to UK ro-ro passenger ships only, but most of the provisions are extended to non-UK ro-ro passenger ships while they are within a UK port by the Merchant Shipping (Closing of Openings in Enclosed Superstructures and in Bulkheads above the Bulkhead Deck) (Application to non-UK ships) Regulations 1988, SI 1988 No. 642, operative 5 April 1988.

Merchant Shipping (Loading and Stability Assessment of Ro-Ro Passenger Ships) Regulation 1989. SI 1989 No. 100

Operative 20 February 1989, the Regulations apply to UK ro-ro passenger ships and in the case of Classes I, II or II(A) require the provision of a loading and stability computer or an equivalent means of making stability calculations and, in general, of an automatic draught gauge system. The Regulations also require records to be made of the ship's draught of water, trim and freeboards and the components of her stability before proceeding on any voyages and for such records to be retained for a specified period.

M1366 should be carefully studied as its main purpose 'is to specify the type of equipment, information and procedures which would be acceptable to the Department in the implementation of the Regulations'. The notice states that the Regulations 'give effect to recommendations by the Court which investigated the loss of the ferry *Herald of Free Enterprise*. Their primary requirements are to ensure that ro-ro passenger ships of Classes I, II, II(A) and IV maintain adequate stability during loading and unloading operations and also that prior to departure their stability is determined in an appropriate manner and shown to be of a requisite standard'.

The owner of every ship shall ensure that information relating to the ship's stability during loading and unloading is included in the ship's stability information booklet.

Officers' duties

1 The master shall ensure that the ship has suitable stability and freeboard.
2 After loading and before proceeding an officer shall ascertain the following information:
   (a) draught at bow and stern
   (b) trim by bow or stern
   (c) vertical distance from waterline to the appropriate loadline mark on each side of the ship.
   The information shall be recorded in the OLB.
3 After loading and before proceeding an officer shall ensure that the KG, GM or deadweight moment (whichever is appropriate) is calculated as per M1366 (a ship or shore computer) and recorded in the OLB. A full record of the calculation shall be retained on the ship for at least one calendar month, and in the case of a Class II or II(A) ship a copy of the record (or the record itself) shall be retained by the owner for at least one calendar month.
4 The master shall cause the maximum permissible KG, or the minimum permissible GM, or the maximum permissible deadweight moment (whichever is appropriate) to be determined and recorded in the OLB.
5 Before a ship proceeds the master shall ensure that the ship has suitable stability standards.

In general, a breach of the Regulations shall be an offence 'punishable on summary conviction by a fine not exceeding the statutory maximum or on conviction on indictment by imprisonment for a term not exceeding two years, or a fine or both'.

However, anyone who provides calculations from a shore-based computer which are not substantially correct is liable on summary conviction to a fine not exceeding the statutory maximum or on conviction on indictment to a fine.

Any officer who is required to ascertain the draught, trim and vertical distance and fails to do so or is careless in carrying out this duty is liable to be punished on summary conviction by a fine of £400 or on conviction on indictment by a fine.

Anyone required to keep a record of stability calculations who fails to do so is liable to be punished on summary conviction by a fine of £400 or on conviction on indictment by a fine.

'All reasonable steps' is a defence. A ship is liable to detention if these Regulations are not complied with.

The requirements of these Regulations are extended to non-UK ships using UK ports by the Merchant Shipping (Loading and Stability Assessment of Ro-Ro Passenger ships) (Non-UK ships) Regulations 1989, SI 1989 No. 567, Operative April 1989.

Merchant Shipping (Emergency Equipment Lockers for Ro-Ro Passenger Ships) Regulations 1988 SI 1988. No. 2272

Operative 1 April 1989, these Regulations apply to UK ro-ro passenger ships. Regulation 4 states:

'4 (1) Every ro-ro passenger ship to which these Regulations apply shall be provided with at least one weathertight emergency locker constructed of steel, or glass reinforced plastic (GRP) or other suitable material at each side of the ship. Such lockers shall contain the equipment specified in M notice 1359. Such equipment shall be of good quality and shall be regularly maintained.

(2) The lockers shall be clearly marked and so located on an open deck and as high up in the ship and as near the ship's side as possible, that in all foreseeable circumstances the locker, or lockers, on at least one side will be accessible'.

M1359 states that 'The provision of such lockers and equipment was a recommendation of the Court of Formal Investigation into the loss of the *Herald of Free Enterprise*'. The purpose of the equipment is to provide means in unusual circumstances, e.g. when a ship is at a very large angle of heel, to assist passengers and crew to escape from enclosed spaces within the ship when the normal escape arrangements cannot be used. Some of the emergency equipment will also be of assistance during damage control operations, and in assisting with escape when the normal escape routes are obstructed by debris or by doors jammed in the closed position'.

M1359 must be closely studied as it gives practical guidance as to the location, construction and equipment of the lockers. In general the equipment should be as follows:

Fireman's axe (long handled)	1
Fireman's axe (short handled)	1
Pin maul (7 lb)	1
Crowbar	1
Hand lamp/torch	4
Padded lifting strop (adult)	6
Padded lifting strop (child)	2
Hand-powered lifting arrangement	3
Lightweight rigid collapsible ladder (at least 3 metres)	1
Lightweight rope ladder (10 metres)	1
First-aid kit	1
Sealed blankets or thermal protective aids	6
Sets of waterproof jackets and trousers	4

All crew members must take part in drills and instruction sessions in all aspects of use of the equipment.

I suggest that the contents of the lockers should be included in the SOLAS monthly checklist of life-saving appliances (Chapter III, Regulation 19 – Operational Readiness, Maintenance and Inspections) and the contents will be examined by a Department of Transport surveyor during the annual survey for the issue of a passenger ship safety certificate. A ship is liable to be detained for non-compliance of these Regulations.

If a ship does not comply with the Regulations the owner and master are liable on summary conviction to a fine not exceeding level 5 on the standard scale, or on conviction on indictment to imprisonment for a term not exceeding two years, or a fine or both. 'All reasonable steps' is a defence.

Merchant Shipping (Operations Book) Regulations 1988, SI 1988 No.1716

Operative 3 April 1989, the Regulations apply to all UK ships of Classes II and II(A), i.e. passenger ships engaged on short international voyages.

Every ship shall be provided by the owner with an 'Operations Book' in which shall be set out instructions to ensure the safe and efficient operation of the ship. The owner shall designate a person who shall be responsible for monitoring the safe and efficient operation of the ship (the Regulations do not go into details regarding who the 'designated' person should be, but someone of the experience and standing of a Fleet Manager or Marine Superintendent would be suitable). The owner shall also ensure means for amending and keeping up to date the Operations Book. The master and designated person shall each keep a copy of the Operations Book and of the Master's Standing Orders (if the later have not been incorporated into the Operations Book). The Operator's Book and Standing Orders (if any) shall be produced on demand to a surveyor of ships.

The master of every ship shall operate his ship in accordance with the instructions in the Operations Book; however, he may deviate from the instructions in the interests of safety.

The owner shall ensure that the designated person is suitable for the duty and is provided with sufficient authority and resources to carry out the duty; contravention is an offence with the penalty on summary conviction of a fine not exceeding the statutory maximum (a) and on conviction on indictment by a fine. Contravention of the other requirements by the owner, master or designated person (as the case may be) can lead to a fine not exceeding £2,000 on summary conviction. 'All reaonable precautions' is a defence. A ship may be detained for non-compliance.

M1353 should be closely studied for guidance in implementing this Regulation. This notice draws attention to M1188, 'Good Ship Management' and the ICS/ISF 'Code of Good Management Practice'. A 'Suggested Outline of

Contents for Operations Book' is contained in the Annex and the following headings are suggested:

Introduction
Shipboard Organization
Shipboard Operations: General
Shipboard Operations: In Port
Preparing for Sea
Shipboard Operations: At Sea
Emergencies and Contingencies

Other relevant M notices

Some other M notices have particular importance for short sea ferry operations.

M1337. This notice draws attention to the Merchant Shipping (Weighing of Goods Vehicles and Other Cargo) Regulations 1988, SI 1988 No.1275 and the Merchant Shipping (Weighing of Goods Vehicles and Other Cargo) (Application to non-UK ships) Regulations 1988 which became operative on 1 February 1989. The purpose of the Regulations is to require owners to ensure that all goods vehicles whose actual or maximum gross weight exceeds 7.5 tonnes, and all other individual cargo items exceeding 7.5 tonnes (except buses) are individually weighed before loading. The Regulations follow a recommendation in the Report of the Court of Formal Investigation into the loss of the *Herald of Free Enterprise* (Report of Court No.8074). That Report pointed out that accurate control over the stability of the ship can only be achieved from a detailed knowledge of the weight and disposition of the cargo. The M notice gives guidance for the operational implementation of the Regulation.

M1299. This notice amplifies and provide general guidance on access opening indicator lights, supplementary emergency lighting and television surveillance systems for ro-ro passenger ships.

M1316. The main purpose of this notice is to draw the attention of all concerned to the relevant statutory requirements on emergency information for passengers.

M1315. Following the *European Gateway* and *Herald of Free Enterprise* casualties, the advice on the stowage of lifejackets in passenger ships has been reviewed and the new guidance is contained in this notice.

M1326. With the introducion of the Merchant Shipping (Closing of Openings in Hulls and in Watertight Bulkheads) Regulations 1987, a greater number of watertight doors will be closed at sea for longer periods. This notice draws attention to the dangers of operating such doors incorrectly.

⸱ *UK Anti-Pollution Regulations*

The Merchant Shipping (Prevention of Oil Pollution) Regulations 1983, SI 1983 No. 1398

Operative 2 October 1983. These regulations have eight parts and eight schedules. The parts are:

1 General.
2 Surveys, Certificates and Oil Record Book.
3 Requirements for Control of Operational Pollution – Control of Discharge of Oil.
4 Requirements for the Segregation of Cargo Oil and Ballast Water.
5 Requirements for Minimizing Oil Pollution from Oil Tankers due to Side and Bottom Damage.
6 Offshore Installations.
7 Reporting of Discharges.
8 Powers to Inspect, Deny Entry, Detention and Penalties.

These regulations, a total of 164 pages, give effect to MARPOL 73 (including Annex 1, Prevention of Pollution by Oil, but no other Annex) as amended by the Protocol of 1978.

Surveys

The United Kingdom has adopted mandatory annual surveys in substitution for unscheduled inspections of all ships requiring an International Oil Pollution Prevention Certificate (IOPP). Annual and intermediate surveys are not required for ships with a United Kingdom Certificate (UKOPP).

The thoroughness of severity of all the above surveys will depend upon the condition of the ship and its equipment. The initial survey should include a thorough and complete examination of a ship and its equipment to ensure that the IOPP/UKOPP Certificate may be issued for the first time. The initial survey will:

1 Examine plans, specifications and technical documentation to verify that the Regulations are being complied with and to confirm that the oil pollution prevention equipment is type approved.
2 Confirm that required Certificates, Oil Record Books, Manuals and other documents are on board.
3 Inspect the construction and installation of the ship and equipment to ensure that all aspects are satisfactory.

The annual survey for an IOPP Certificate must be held within three months before or after the anniversary date of the IOPP Certificate. In general, the annual survey should consist of:

(a) an examination of Certificates, a visual examination of a sufficient extent of the ship and equipment (and certain tests) to confirm that everything is being properly maintained; and

(b) a visual examination to confirm that no unapproved modifications have been made.

The UKOPP and IOPP Certificates are valid for a period of five years. The initial and any renewal surveys must be carried out in accordance with the detailed procedures specified in M1076.

* *Oil Record Book*

Most ships must be provided with an 'Oil Record Book Part 1 (Machinery Space Operations)', and every oil tanker of 150 GRT and above shall be provided with an 'Oil Record Book Part II (Cargo/Ballast Operations)'. The book must be completed whenever any of the following operations takes place:

(a) Machinery space operations in all ships
    1 ballasting or cleaning of oil fuel tanks
    2 discharge of ballast or cleaning water from (1)
    3 disposal of oily residues such as sludge
    4 overboard discharge of machinery space bilge water

(b) Cargo/ballast operations on oil tankers
    1 loading of oil cargo
    2 internal transfer of oil cargo during a voyage
    3 discharging of oil cargo
    4 ballasting of cargo tanks and dedicated clean ballast tanks
    5 cleaning of cargo tanks including crude oil washing
    6 discharge of ballast except from segregated ballast tanks
    7 discharge of water from slop tanks
    8 closing of slop tank valves after discharge operations
    9 closing of isolation valves to dedicated clean ballast tanks
    10 disposal of residues

The term, 'dedicated clean ballast tanks' means tanks that are dedicated solely to the carriage of clean ballast, they are an alternative to segregated ballast tanks but the regulations concerning their application are particularly tortuous.

In the event of any oil discharge into the sea for the purpose of securing the safety of a ship or saving life or in the event of a discharge due to an accident or exceptional circumstances, a statement must be made in the Oil Record Book regarding the incident.

Details of each operation should be put in the book as soon as the operation is completed and the entry signed; each completed page should be signed by the master. The book should be kept readily available for inspection by any authorized person and it should be kept for three years after completion. It should be noted that the Department of Transport has given authority to the Harbour Master, or any other person employed by a Harbour Authority, to inspect the Oil Record Book of a ship in a British harbour and to make a copy of any entry in the book.

Discharge of Oil or Oily Mixture into the Sea by Tanker

Such a discharge can only be carried out if the following conditions are complied with;

1 The tanker is proceeding on a voyage.
2 The tanker is not within a special area (basically the Mediterranean, Baltic, Black and Red Seas and the Gulf area).
3 The tanker is more than 50 miles from the nearest land.
4 The instantaneous rate of discharge of oil content does not exceed 60 litres per mile.
5 the total quantity of oil discharged does not exceed 1/30,000 (new ships) or 1/15,000 (existing ships) of the particular cargo of which the residue formed a part.
6 The tanker has in operation an oil discharge monitoring and control system plus an approved slop tank arrangement as stipulated by Regulation 15.

Oil Discharge Monitoring and Control Systems for Retention of Oil on Board

The details of these systems are contained in Schedule 4 and should be studied in full. The system shall record continuously:
(a) The discharge of oil in litres per mile and the total quantity of oil discharged; *or*
(b) instead of the total quantity of oil discharged, the oil content of the effluent and rate of discharge.

The record shall be 'indentifiable as to the time and date' and must be kept for at least three years. The system shall come into operation when there is any discharge of effluent into the sea and must automatically stop any discharge when the rate of discharge of oil exceeds 60 litres per mile.

The specifications for oily-water separating equipment and oil content meters are given in Schedule 3. Separating equipment must be capable of producing an effluent containing not more than 100 ppm of oil irrespective of the oil content of the feed supplied to it. Filtering equipment must be capable of reducing the oil content in the effluent to not more than 15 ppm. Oil content meters must be capable of measuring a range of oil content and an alarm must be provided which will indicate when the oil content of the effluent exceeds

15 ppm. The figure of 15 ppm is used to determine whether ballast is clean or not. Rule 1(2) states that if ballast is discharged through an approved system and that if the oil content of the effluent does not exceed 15 ppm that figure 'shall be determinative that the ballast was clean'.

In general, the system will be designed so that no ballast discharge overboard can take place unless the monitor is operating normally. The system should have a minimum number of discharge outlets and the sampling points (for monitoring purposes) should be so arranged that discharge can only take place via only one sampling point at a time.

A system must comprise the following:

an oil content meter
a flow rate system to indicate the quality of the effluent per unit time
a device to give the vessel's speed in knots
a sampling system to convey a representative sample of the effluent to the oil content meter
a control unit to process signals, activate alarms and operate recorders with means to prevent effluent discharge before the meter is working and with a manual override system.

Reporting of Discharges

Although the regulation specifically applies to British ships it can be said that in general reports must be made:
(a) by all ships within 200 miles of land
(b) by all oil tankers when fully or partly loaded
(c) by all ships of 10,000 GRT and above

The master must make a report whenever 'an incident involves any discharge or probable discharge of oil or oily mixtures as a result of damage to the ship or its equipment or for the purpose of securing the safety of the ship or saving life at sea'. The types of incident include collision, grounding, fire, explosion, structural failure, flooding, cargo shifting, and any breakdown which impairs steering gear, propulsion plant or generators or which impairs the safety of navigation.

The report should, in general, be made to the nearest coastal State via coastal radio stations. It should contain the following particulars:

1 Name of ship, call sign, frequency or radio channel kept open.
2 Date and time of incident.
3 Position and extent of pollution including, if possible, estimated amount and surface area of spill.
4 Present position of vessel if different from (3).
5 The rate of release if the discharge is continuing.

6 Wind direction and speed, and the condition of any current or tide affecting the spill movement.
7 General weather conditions and the sea state at the ship's present condition.
8 Type of oil discharged.
9 Types and quantities of all oils still on board (including the ship's fuel).
10 Type of ship, size, nationality and port of registry.
11 Course, speed and destination if still proceeding on voyage.
12 Brief description of the incident, including any damage sustained and the cause of any discharge.
13 Ability to transfer cargo or ballast or bunkers.
14 Any action being taken by the ship.
15 Forecast of likely movement and effect of pollution with estimated timing.
16 Assistance which has been requested or which has been provided by other ships or agencies.

After the initial report a further updating report must be sent to the authority which received the first report.

Ships of less than 400 GRT (other than oil tankers)

M1240 should be read for this information as the purpose of the notice explains the extent to which the Merchant Shipping (Prevention of Oil Pollution) Regulation 1983, as amended, applies to the above vessels.

Inspections

It should be noted that these Regulations apply to United Kingdom ships and non-British ships in UK waters. Regulation 32 states that inspections can be carried out to verify 'that there is on board a valid IOPP Certificate in the form prescribed by the Convention or UKOPP Certificate in a form prescribed in Schedule 1'.

A ship can be denied entry or be detained if these Regulations are not fully complied with.

Penalties

The penalties can be divided into two basic 'categories'.
(a) If any ship puts effluent over the side which can cause pollution the owner and the master 'shall be liable on summary conviction to a fine not exceeding £50,000 or on conviction on indictment to a fine'.
(b) Failure to comply with the other requirments of these Regulations shall make the owner and the Master liable to a fine not exceeding £1,000 on summary conviction or 'on conviction on indictment by a fine'.

'All reasonable precautions' and 'all due diligence' are a defence.

The Prevention of Pollution (Reception Facilities) Order 1984, SI 1984 No. 862

Operative 25 July 1984, this Order gave effect to the provisions of regulation 12 of Annex I and Regulation 7 of Annex II of Marpol 73/78. It requires harbour authorities and terminal operators to provide reception facilities for ships which are using the harbour or terminal for a primary purpose other than using the reception facilities. The facilities must be adequate to meet the needs of ships using them without causing undue delay to ships. M1338 gives very comprehensive instructions and guidelines for the practical application of this Order. Oily residues reception facilities have been required since 2 October 1984 and noxious liquid substances residues and mixture reception facilities have been required since 6 April 1987.

Part of the instructions are based on the IMO publication *Guidelines on the Provision of Adequate Reception Facilities in Ports; Parts I and II.*

Over The Side Is Over!
This was the title of a UK Department of Transport poster to draw owners' attention to new regulations which brought Marpol Annex V into force in the UK.

Merchant Shipping (Prevention of Pollution by Garbage) Regulations 1988. SI 1988 No. 2292

Operative 31 December 1988, these Regulations give effect to Regulation 1–6 of Annex V and apply to United Kingdom ships anywhere and to non-British ships in UK waters. The Regulations stipulate the requirements as to the disposal of garbage from ships into the sea outside Special Areas and the more stringent requirements as to disposal within Special Areas.

Merchant Shipping (Reception Facilities For Garbage) Regualtions 1988. SI 1988 No. 2293

Operative 31 December 1988, these Regulations give effect to provisions of Annex V with regard to garbage reception facilities and they apply to every harbour authority or terminal operator in the United Kingdom. The Regulations empower harbour authorities and terminal operators to provide reception facilities for garbage from ships and require such facilites to be adequate. Reasonable charges can be made for the use of such facilities.

Marine Pollution Statutory Instruments

In May 1989 there were twenty-five statutory instruments relating to marine pollution. Thus ship's officers and shore managers have an onerous task to ensure that the regulations are being complied with. It is therefore important that all tanker management staff are conversant with Marpol 73/78 and the national legislation which implements it for individual countries – British regulations can be accepted as being typical of such national legislation. The continuous flow of regulations does make it difficult for management to have the required level of legislative knowledge for efficient operation. I can only point you in the right direction; in addition to the above, mangers should have knowledge of:

The Merchant Shipping (Control of Pollution by Noxious Liquid Substances in Bulk) Regulations, SI 1987 No.551

The Merchant Shipping (Reporting of Pollution Incidents) Regulations, S.I. 1987 No. 586

Ship and shore management should keep an up-dated file of M Notices. A particluar notice which lists the 'Principal Acts and Regulations on Merchant Shipping' is issued at regular intervals and that notice should be up-dated when new regulations come into force.

3

Ship Maintenance: Corrosion Prevention

Corrosion

There are many definitions of corrosion but seamen can consider it as the deterioration of a metal due to an electrochemical reaction with its environment.

Most metals are unstable and have a tendency to return to their natural state. Such metals tend to react with their environment and produce compounds, e.g. iron oxide or 'rust'. The process involves the movement of electrons and it is this electrochemical reaction which causes the main problems of corrosion on board ship.

If two electrically connected metals, each of a different potential, are immersed in a common electrolyte, electrons will flow from one metal to the other. The two metals are electrodes. The one which loses electrons is known as the anode and the one which receives them, the cathode. This combination of metals, electrolyte, and electron flow is known as a galvanic cell and is shown in Figure 3.1

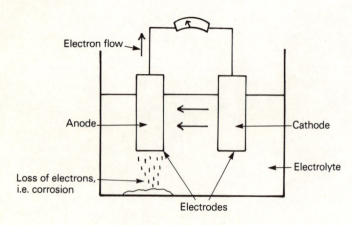

Figure 3.1   Galvanic cell

The loss of electrons converts a metal into its compounds. Thus the anode corrodes while the cathode remains unchanged. Corrosion on board ship means the loss of electrons from metallic structures to the environment, i.e. water and air, with the result that oxides (usually 'rust') form on the metal surfaces.

A concentration corrosion cell is shown in Figure 3.2. The break in mill scale or a paint film exposes a small portion of the ship's plate to the atmosphere; thus an electron flow occurs. The unbroken surface acts as the cathode and the electrons flow away from the break in the paint film towards the cathodic area. Thus corrosion occurs at the exposed metal, the metal being the anodic portion of the cell. This process is known to mariners as 'pitting'. If all the ship's metal surfaces were left unprotected, corrosion would proceed at a relatively uniform rate all over the ship; the main engines would fall to pieces before the ship would rust away. However, for various reasons we paint the ship's structure and we must therefore prevent the formation of corrosion cells.

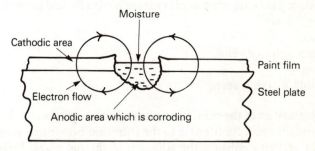

Figure 3.2   Concentration corrosion cell, caused by a break in the paint film

Corrosion should not be confused with erosion, which is the destruction of material by the mechanical movement of liquid or gas. The effect of erosion is increased if solid particles of matter are contained in the medium flowing though a pipeline. Crude oil which contains impurities will cause internal erosion in a pipeline quicker than refined oils. Turbulence at sharp bends adds to the erosion process and most pipeline leaks occur at such locations. Abrasion will remove protective coatings and expose metal surfaces to the atmosphere, thereby creating corrosion cells. Most of the ship's side rusting is caused by abrasion/corrosion.

The rate of 'rusting' varies with humidity. Unfortunately for the mariner the salt-laden humid atmosphere which surrounds his vessel is more conducive to rust than any other environment on earth. Corrosion will not occur unless oxygen and water are both present. It may be of some use to consider the process as a 'corrosion triangle' similar to the fire triangle:

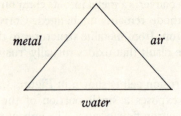

Corrosion will not occur unless both water and oxygen are present. Thus if we can exclude one, we can prevent the formation of corrosion cells. The most convenient method the mariner has for excluding both air and water is *paint*.

Basic composition of paint

Some modern paints are of a complex nature but traditional paints have three basic components:
(a) pigment
(b) binding agent or vehicle
(c) solvent
and possibly a drying agent.

(a) The *pigment* gives the paint its colour and covering capacity. In a primer it should be the main contributor to the corrosion-preventing properties of such paint and also enhance the adhesion of the top coats. Pigments are generally dry powders which are held in suspension in the paint and vary from natural minerals to man-made organic compounds. They can be divided broadly into:

    colour pigments such as
        white—white lead, zinc oxide, titanium dioxide
        black—carbon
    primers such as red lead, zinc chromate, calcium plumbate
    metals such as aluminium and zinc

The pigment is held in suspension in a solution of a binding medium.

(b) The *binding medium* is the most important component of paint and as well as determining its consistency and application gives the paint its physical and chemical properties. The possible variation in binder properties is limitless and the student might be wise only to consider the two basic types of oleoresinous and alkyd binders.

    The oleoresinous vehicle combines drying oils with natural and some synthetic resins.

The alkyd vehicle combines drying oils with synthetic resins based on alcohols and acids.

Some of the basic ingredients found in binders are given below.

Drying oils—linseed, tung, soya bean

Natural resins—copal, damar, rosin

Synthetic resins—phenolic, vinyl, epoxy, polyurethane, silicone

(c) The *solvent* is the volatile component of paint and its function is to make the paint flow for ease of application. It is generally the action of the solvent evaporating that makes paint adhere to a surface.

Modern paint types will be discussed in a later section.

Candidates for Certificate of Proficiency/HND Nautical Science, when answering examination questions on corrosion protection, should remember that paint is still the most important protection system for ships. A general answer on anti-corrosion systems should commence with a discussion on paint before moving on to describe other methods.

Cathodic protection

Corrosion is difficult to control on underwater areas of the ship's structure as inspection presents problems and even when corrosion is detected corrective action would sometimes be impossible. All corrosion is basically 'galvanic' but most mariners use the term 'galvanic corrosion' when discussing the problem of dissimilar metals being located close to one another when both are in an electrolyte. The corrosion rates of metals and alloys which are submerged in sea water vary considerably and extensive research has been conducted on the problem. Metals which corrode rapidly are known as anodic or ignoble metals and those which resist corrosion are termed cathodic or noble. In Figure 3.3 the ignoble metal, steel, is corroding while the noble

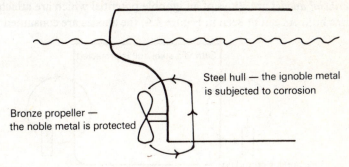

Figure 3.3   Galvanic corrosion. The main area of corrosion is within the electron flow line

metal bronze is protective. The corrosion rates of metals are shown in tables which are known as 'Galvanic Series': Table 3.1 considers metals and alloys which are commonly used in shipbuilding.

Table 3.1   Metals and alloys commonly used in shipbuilding

Ignoble or anodic end (corroding)	Magnesium
	Zinc
	Aluminium
	Iron, steel
	Lead, tin
	Nickel
	Mill scale
	Copper
	High duty bronzes
	Stainless steels
	Titanium
Noble or cathodic end (protected)	Platinum

Any metal in the above series will therefore be subjected to accelerated corrosion when located in an adjacent position to a more noble metal. The underwater areas most likely to be affected by this process are:

(a) the stern region due to the bronze propeller
(b) the vicinity of engine room inlets and discharges
(c) valve fittings in tanks

The most efficient system for combating underwater corrosion is 'cathodic protection'. The basic principle of this method is that the ship's structure is made cathodic, i.e. the anodic (corrosion) reactions are suppressed by the application of an opposing current and the ship is thereby protected. Cathodic protection, which is only possible when metals are immersed in an electrolyte, is provided by two systems (1) sacrificial anodes, and (2) impressed current.

1 *Sacrificial anodes* are alloys of an ignoble potential which are attached to the ship's hull. As can be seen in Figure 3.4, the anodes are consumed while

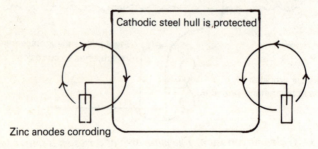

Figure 3.4   Sacrificial anodes

the metal is protected, the flow of electrons being from the anodes to the ship's structure. This method can be used in ship's tanks, especially the cargo and ballast spaces of tankers, or on the hull. As the anodes supply the power, there is no costly outlay for equipment, the installation is simple, no supervision is required, and the current cannot be supplied in the wrong direction. The effectiveness of the method depends on the correct current flow being maintained and a formula is used to calculate the amount of anode required for a given protection area for a given time. If the estimation is incorrect the anodes may dissolve completely between drydocks or repair periods and protection is lost. It is therefore essential that anodes in cargo and ballast tanks are inspected at regular intervals and a record of weardown kept. If possible, a few unused anodes should be kept in a convenient location so that weardown comparisons can be made easily. The securing arrangements of tank anodes should be checked as vibration sometimes loosens them. Hull anodes should also be checked for weardown, contact damage, vibration damage, and to check that no overenthusiastic sailor has painted them. The turbulence around the stern could cause uneven weardown and the increased oxygen supply provided by turbulence could accelerate the wastage rate. Sacrificial anodes are made of alloys of zinc, aluminium, or magnesium but the latter should not be used in cargo tanks because of the 'spark hazard' which could be caused by one falling and hitting the tank structure.

The effectiveness of galvanic protection depends on the current flow and with anodes the current available depends on the anode area. The number required to protect the hull area of large ships could cause too much turbulence, the total weight would be excessive, and the anodes would be too costly and cumbersome to fit. The impressed current system would therefore be used.

2   *Impressed current systems* can only be used to protect the immersed external hull. Anodes are fitted to the hull which is protected from corrosion by maintaining a voltage difference between the anodes and the hull. An AC current from the ship's electrical system is fed into a rectifier and DC power is supplied to the anodes. A silver/silver chloride reference cell on the hull measures the current density in the sea water, i.e. the voltage difference between itself and the hull. The reference cell indirectly regulates the power to the anodes by means of a controller which amplifies the micro-range reference cell current and compares it with a predetermined fixed potential difference. The 'difference' between the two readings is fed back to the rectifier which then alters the current being supplied to the anodes until the predetermined and reference cell potentials are equal. Figure 3.5 shows an impressed current system.

A potential difference within the range of 180–250 mV would be suitable for most ships but current density tables relating to the area of the wetted hull

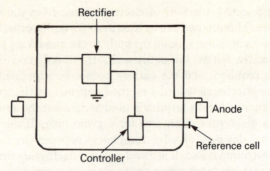

Figure 3.5 Impressed current system

are used to determine the number of anodes and the potential difference required to protect a vessel. Non-consumable relatively noble metals such as lead/silver are used for the anodes which are cast in fibre-glass bodies. The bodies are bolted to flat bars which are welded to the shell plating, the cables from the anodes being led through the hull via special ducts which are built to stringent construction requirements. The anodes are insulated from the hull by shields which are fitted to prevent high-density currents stripping the hull paint. Due to the size of VLCCs (very large crude carriers) a second power unit is located forward but the anodes for that unit must be recessed to prevent them being damaged by the anchors. Three factors determine the position of the anodes:

1 They must be located where there is the least risk of contact damage.
2 They must interfere as little as possible with the water flow around the ship and to the propeller.
3 The correct current density must be maintained.

The impressed current system is more expensive to install than sacrificial anodes but as the latter must be renewed, the eventual cost of both systems would appear to be similar. The impressed current method requires trained personnel to supervise the power units. Log sheets containing the daily readings of volts, millivolts, and the current in amps must be maintained and the reference reading of millivolts should be compared with the predetermined value. Although the units are automatic, they must be put on manual control when alongside in port as the system may attempt to protect the jetty structure. The anodes and reference cells must be inspected frequently for damage and also to ensure that they have not been painted over.

Bimetallic corrosion

Strictly speaking, any two different metals which are in electrical contact and are bridged by an electrolyte are liable to bimetallic corrosion. Thus, such corrosion will occur between metals which are not in direct contact with each other. However, many mariners use the term to describe the reaction which occurs when two different metals are in direct contact. Many corrosion and breakdown problems occur on board ship because metallic parts have been coupled without adequate insulation or protection being provided. The rate of corrosion in the ignoble metal will depend upon the relative sizes of the anodic and cathodic materials. A small anode attached to a large cathode, e.g. a steel flange in a copper pipe, will result in rapid corrosion of the anode. If design requirements are such that two dissimilar metals are in close proximity to one another the following procedures should be observed:

1 The more noble metal should be used for actual connections.
2 Both metals should be covered by a paint film to a very high standard.
3 The metals must be efficiently insulated from each other.

Areas liable to bimetallic corrosion are:

1 Valve fittings in tanks.
2 Aluminium superstructures attached to steel decks.

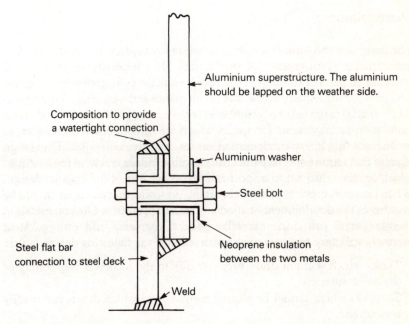

Figure 3.6 Aluminium and steel bimetallic coupling. All surfaces are in tight contact with one another but to show clearly the neoprene insulation the area around the steel bolt has been expanded

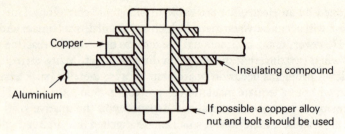

Figure 3.7   Aluminium and copper bimetallic coupling

3 Steel lifting hooks in aluminium lifeboats.
4 Openings in superstructures, such as portholes and windows.
5 Inlets and outlets to tanks and to engine-room piping systems.
6 Zinc and aluminium drainage channels.
7 Bolted joints.

Figures 3.6 and 3.7 show the insulation arrangements at two bimetallic couplings.

Design faults

The designers and builders of ships occasionally neglect the operational and maintenance requirements of ship's staff. The necessity to inspect and maintain all parts of the ship's structure is sometimes forgotten by designers who provide inadequate access and maintenance arrangements. Void spaces under special cargo tanks or chambers are particularly difficult to work in and ventilation arrangements for spaces which are seldom entered are often so inadequate that heavy condensation causes excessive corrosion. The design of areas that require special paint coatings often induce cracks in the coatings, which are then difficult to inspect and repair because of the original design. When possible, crevices or any other moisture-collecting locations should be avoided as the development of anodic breaks in protective films on metals in stagnant areas can cause excessive general corrosion and pitting. Most mariners will have had to contend with some of the following design faults:

1 Tanks which will not drain properly due to the incorrect positioning of drains or suctions.
2 Scuppers which cannot be cleared easily and in which debris can readily accumulate.
3 Pipelines with excessively sharp bends which cause erosion/corrosion.
4 Channels and gutters without drain holes.
5 Inaccessible areas which cannot be painted.

6 Lagging which is a moisture trap.
7 Areas under deck machinery and equipment with inadequate drainage facilities.
8 Incompatible bimetallic couplings.

Plate preparation during building and repair periods

The initial surface preparation of new plate, whether for building or repair purposes, is of paramount importance as it determines the life of the protective paint system which is subsequently applied and affects the maintenance of the paint coatings. In the case of underwater hull plating it can affect the roughness of the hull, which is one of the determining factors of the ship's speed, fuel consumption and operational costs.

The use of high-performance synthetic resin paints requires a very high standard of plate preparation. As well as being very expensive, such paints do not have the tolerance of poor surface preparation which was one of the advantages of oleoresinous paints. All contaminants such as grease, dirt, salts, atmospheric pollutants, and rust and mill scale must be removed from the plates prior to painting.

Mill scale

This is a bluish-black iron oxide layer which forms on steel plates during hot rolling at the mill. The scale is very corrosion resistant and the anodic process of corrosion cannot take place where the scale is pore-free and adherent. However, the mill scale will not be continuous over a surface, fissures will appear at welded seams, and pieces of the scale will become detached during handling. Severe corrosion will occur at those places as the anodic reaction is only possible in the areas where the steel is exposed. The scale is conductive and the cathodic process (the consumption of oxygen) will take place readily over the whole unbroken surface. As the total amount of corrosion will be determined by the consumption of oxygen on the entire surface, intensive and rapid pitting can occur. Until the early 1960s mill scale was removed by the process of weathering. Shipyards stockpiled large amounts of plate which were left in open stockyards for several months so that atmospheric exposure caused rusting which pushed the scale off the metal. This method is no longer satisfactory and other preparation techniques must be used.

Flame cleaning

Mill scale and steel have different coefficients of expansion. Thus when steel is heated the scale flakes off. The plate surface is rapidly heated by an

111

oxyacetylene flame and the loosened scale/rust mixture is brushed off. Steel can be damaged by rapid heating and the process has to be controlled carefully and monitored to ensure that the temperature does not exceed 150°C. The principal advantage of this method is that it leaves the surface absolutely dry and primer can be applied when the plate is still warm, around 30–40°C. This advantage is usually lost as the process is not totally effective and residue rust/scale deposits still require to be removed by other methods. Although there is little health hazard associated with flame cleaning, exposure to the high temperature and flames must be avoided. To cover large areas multi-flame heads are required and flame cleaning cannot be carried out on certain vessels, e.g. tankers in repair yards.

Acid pickling

The handling process can be difficult and it is not suitable for shipyards which have a large throughput of heavy steel plates. The method is sometimes used for individual plates and small complicated sections. It is ideal for the preparation of thin plate used in the construction of smaller craft. Sulphuric and hydrochloric acid solutions are generally used but the solutions must also contain a wetting agent and an inhibitor to reduce metal loss. After pickling it is essential that the plates receive a thorough rinse in fresh water. One of the disadvantages of this system is that it leaves the plates wet. However, a method has been devised in which hot liquids are used and a coat of primer can be applied when the plate is still warm. Dissimilar steels should be pickled separately.

Blast cleaning

This method is now standard practice in all major shipyards and is generally considered to be the best way of removing mill scale, rust and most contaminants. Most units consist of a heater or drier, a blaster, and an automatic paint sprayer. One disadvantage of the system is that the blasting 'activates' the steel, making it highly susceptible to the corrosion process. Thus the primer must be applied as soon as possible and only small areas can be cleaned at any one time. Large fixed plants are capable of cleaning 500 m²/hour but this rate is severely reduced for ship's side dry dock operations.

Removal of the abrasive can be a problem. In modern units the abrasive, scale and rust falls into a container from which it is removed to a separator. The abrasive is then re-used but this process, plus the maintenance of the entire unit, add to the cost of the operation.

Steel shot and grit are two abrasives which are widely used. Round shot can hammer mill scale into the plate causing peening; this may be defined as being the deposition in powder form on a plate of a coating metal, in this case mill scale. Successive shot blasting can then build up a work hardened surface which curls up into hackles. These miniature ridges can be difficult to cover with paint and may form rust spots. Grit is recommended by some shipyards as it chips the corrosion off, rather than hammering it into the surface, the plate is only slightly roughened and this improves the paint adhesion.

Thin plate can be distorted by blasting and the use of the wrong size of grit can produce an inferior surface. The operators of blast plants must be highly skilled as the cleaning process varies with the condition of the surface being treated. The dust produced by blast cleaning is a health hazard and suitable precautions must be taken to eliminate that hazard. The use of sand may cause silicosis but it is no longer used in the United Kingdom.

Causes of paint failure

Mariners are engaged in a constant battle against corrosion and often feel that they are on the losing side in a war of attrition. The ocean is a relentless enemy and the salt-laden elements constantly expose weaknesses in paint systems in their quest to turn steel into rust. Many factors contribute to paint failure and experienced mariners will probably be able to add to the following list:

1 Poor surface preparation.
2 Painting in an unfavourable environment. As the ideal conditions fall within a range of 10°C and 32°C and a relative humidity below 90 percent mariners will seldom find themselves in ideal conditions.
3 Underestimating the corrosive nature of sea air and not applying enough paint coats.
4 Using paint which is too thick or too thin.
5 Incorrect mixing of two-pack paints.
6 Prolonged exposure to the environment, e.g. engine fumes.
7 Expansion and contraction of the metal surface which loosens inelastic paint.
8 Excessively high temperature which causes cracking.
9 Paint application by incorrect methods, e.g. wrong nozzle size on a spray gun or using a roller for paint which must be applied solely by a spray gun.
10 The use of incompatible coats, e.g. an incorrect primer for a following coat.
11 The use of unsuitable paints in specific areas.
12 Poor design, e.g. areas difficult to paint.
13 The use of unsuitable barrier paints beneath anti-fouling.
14 Chipping of paint surfaces during normal working operations.
15 Abrasion due to the vessel landing heavily on the quayside.
16 Insufficient drying time between coats.

Shipboard preparations for painting

If the paint surface is undamaged, only the following steps are necessary.

1 If the surface is contaminated, a solvent cleaner should be applied.
2 The surface should then be washed down with a warm detergent solution and rinsed off with a copious amount of fresh water. It is essential that all traces of salt are removed from the old paintwork.
3 The surface must be thoroughly dried before painting commences.
4 If the previous paint still retains a hard gloss texture, it should be roughened and an undercoat applied.

Epoxy paint is difficult to 'touch-up' and it may be necessary to recoat entire plates, the appropriate undercoating being first applied.

If the paint surface has broken down to a significant degree then the complete surface should be cleaned to bare metal. The chipping hammer and wire brushing method does not produce a satisfactory surface for the adhesion of modern paints. The use of hand hammers invariably produces tiny ridges in the steel and thus increases considerably the area of steelwork which is vulnerable to corrosion. The chipping also produces small hollows which are difficult to cover effectively with paint and which become moisture traps. Chipping can therefore be a prime factor in initiating the pitting process. Use should be made of modern methods of surface preparation.

Power wire brushing

A rotary wire brush is driven by electricity or compressed air and the brush, which is circular and usually hand held, produces a good surface when used against moderate rust patches.

Power discing

A similar unit to the above, but the head is a rotary abrasive impregnated nylon disc or a backing disc with detachable abrasive paper. The power disc is probably the slowest cleaning method but it produces a good surface on small moderately rusted areas. It can be used to good effect when roughening epoxy paint surfaces to provide better adhesion when touching up.

Air hammer

Powered by compressed air, some 'windy hammers' are fitted with reciprocating heads, the size of a small hammer. When cleaning irregularly shaped surfaces, a set of needles can be inserted instead of the hammer, but the needles do have a tendency to break. The cleaning rate is moderate but it can be used effectively against heavy scale and the needles can clean areas of pitting. Operators should wear ear protectors as the noise produced can be considerable and the sleeping requirements of watch-keepers should be kept in mind.

High-pressure water blasting

A sophisticated and increasingly popular method of rust removal, it produces a 'white metal' finish suitable for modern paint. One leading manufacturer uses a diesel engine to drive ultra-high pressure pumps capable of producing a pressure of 10,000 psi and a water flow of 10 gallons a minute. The water flows from the pump by a hose and is blasted out through a hand-held pistol-grip gun which directs the jet directly against the surface to be cleaned. The gun nozzles can be changed by hand and special purpose guns can be used for difficult areas and pipe cleaning. The guns are fitted with fail-safe triggers.

The power unit which has wheels or skids can be moved about the ship and the method is very effective for descaling decks. Sand can be injected with the water to increase the efficiency of the operation. Water blasting is perhaps the fastest of the 'on board' descaling methods and it produces excellent surfaces.

Safety precautions should be observed when using all power equipment and safety hats, goggles, gloves and safety shoes should be used by operators.

Modern paint types

Alkyd

The vehicle in this type of paint is based on alcohols and acids. It gives an improved drying time over vegetable oils; thus alkyds are mainly finishing paints suitable for surfaces exposed to the atmosphere. The performance is poor if the surface is immersed. Alkyd paints should only be used on superstructures and not under water.

Bitumen or pitch

The paint is made by dissolving bitumen or pitch in solvents such as naphtha and white spirit and may be used as a superstructure paint but more commonly for internal surfaces exposed to a high degree of wetness such as fresh-water tanks. When pigmented with aluminium paint it is often used for plating which is constantly under water and for outer shell plating.

Chlorinated rubber

This paint is used where good chemical and water resistant coatings are

115

required. Normally single pack it consists of pigments of plasticized chlorinated rubbers. The paint dries by the evaporation of the solvents and the temperature at the time of drying is not as critical as with some other paint types. It is often used when ships are being built in winter or in poor climatic conditions. It is designed for use in high-build systems, i.e. intended for application with airless spray which gives very high film thickness (80–100 microns) and substantially reduces the number of coats required. This type of paint is particularly resistant to acids and alkalis and is often used to protect outer shell plating. Maintenance is assisted by its good adhesion to previous coats of the same material but it cannot be applied to conventional paint surfaces. This paint can also be used for superstructures but in some cases cracking has occurred after a few years.

Coaltar epoxy

A two-pack paint in which the components are pigments (bitumen or coaltar pitch blended with resin), epoxide resins, and curing agents. The paint provides hard thick films in which the chemical-resistant qualities of epoxy resin are combined with the water impermeability of coaltar. It is widely used in immersed conditions but it can be susceptible to atmospheric corrosion and is easily damaged.

Epoxy

Epoxy resins are derived from chemicals which are extracted from petroleum and natural gases. The epoxy resin paints have extremely good chemical, water, and abrasion resistant qualities but are very expensive. They are generally two-pack (two components) consisting of an epoxy base and a hardener or curing agent. Drying is caused by the chemical action of the polyamine or polyamide hardener with the epoxy resin base, i.e. the paint is of the reaction-drying type. The paints are very hard when cured and give long-term protection against corrosion. Their mechanical resistance properties are superior to chlorinated rubber and vinyl paints. Initial drying, which is by solvent evaporation, takes place in between four and six hours, but this is followed by the chemical cure of the binder, the rate of which is dependent on the temperature. Drying and hardening will only occur satisfactorily in temperatures above 10°C and in relative humidities below 80 percent. One often has to wait for considerable periods of time on board ship before suitable weather conditions are available for the application of this type of paint. Due to the expense involved, and the short pot life, only sufficient paint for a half day's use should be mixed at any one time. The penetration quality is rather poor and first-class surface preparation is essential to obtain a good performance from this paint, touching up is difficult and the previous coat must be roughened. Gloss finish tends to 'chalk' and epoxy is to be preferred as a matt deck paint than as a decorative bulkhead paint. Epoxy paints are difficult to maintain but if applied under suitable conditions the performance is superlative.

Oleoresinous

An improvement on oil-based paints, the vehicle consists of drying oil and natural or synthetic resins, including phenolic resins, which are often synthesized from several different oils and resins. These paints have improved drying time and give a better performance than oil-based paints but are not as sophisticated as the chemical-resistant paints.

Phenolic

Phenol is derived from coaltar and produces a water-resistant paint. Paints produced from 100 percent phenolic resins have excellent chemical resistance and a very hard finish. Although often used to coat tanks on chemical carriers, unmodified phenolic paints are sometimes used on weather deck plating and give good corrosion resistance. Modified phenolic paints are often used in order to reduce costs.

Polyurethane

Usually a two-pack paint, an alkyd polyester resin is mixed with an isocyanate hardener and the ensuing reaction produces polyurethane paint. These paints have many good qualities, abrasion resistance, hardness, a high gloss, chemical resistance, and when well cured have very good water and weather resistance. Polyurethane paints are used for tanks coatings but white polyurethane is good external superstructure paint as it remains bright for long periods and is not subject to rapid yellowing. This type of paint is approximately four times the cost of alkyd paints but the superior performance and reduced maintenance costs make it attractive to mariners. Unfortunately the paint is very sensitive to moisture and high humidity during application and can be difficult to use in the marine environment. The hardness does cause a tendency to chip or crack but the advantages of this paint vastly outweigh its disadvantages.

Primers

Steel plate must be protected from corrosion while the ship is being built and prefabrication primers must also be rapid drying, have a non-toxic vapour, not affect weld quality or speed, be suitable for spray application, and be a suitable first coat for the ship's paint system. The most successful anti-corrosive primers are those which are pigmented with iron oxide or zinc dust and which contain corrosion inhibitors such as zinc chromate or zinc phosphate. Zinc dust silicate paints are used both as prefabrication primers and as long-term anti-corrosive paints for dry cargo holds, cargo tanks, and ballast tanks. Such coatings resist high humidity, high temperatures, condensation and abrasion, thus are very suitable for use in marine environments. The paints have exceptional durability and good protective properties and are frequently used in tanks which carry latex, and alkalies. Zinc silicate paints can be overcoated with finishing paints to give increased performance.

The most effective primers which the mariner can use for routine maintenance are those which contain chemical corrosion inhibiting pigments. Calcium plumbate is a good alternative to red lead. When used with an alkyd base it is quick drying and has excellent adhesion. The major disadvantage is that it has toxic properties. One of the best primers available is zinc chromate which is particularly suitable for use on zinc, galvanized steel, aluminium and other non-ferrous metals. It is suitable for brush or spray application and is non-toxic.

Mariners must ensure that any primers used are compatible with following coats, e.g. a vinyl primer must be used underneath a vinyl topcoat. Zinc-rich two-pack primers include vinyl, methane, polyester, silicote, and alkyd paints. One of the most popular types is a zinc-rich epoxy primer which is rapid drying and has excellent corrosion resistance. One-pack zinc-rich polyurethane primers give good performance and, although mainly used on ship's machinery, may become increasingly popular due to easy application.

Vinyl

The resins are produced by the polymerization (the combining of small molecules to form large molecules) of organic compounds in the vinyl group. The paints have a low solids content due to the relatively low solubility of the resins and this results in thin dry films. More coats are therefore required to build up an adequate paint thickness and this increases the application costs. Adhesion is poor if surface moisture is present; therefore good weather is required during the application and because of this the paint is not popular in the United Kingdom. Adhesion to bare steel is also poor and the paint must be applied over a pre-treatment primer such as zinc chromate in a polyvinylbutyral resin which uses phosphoric acid as an etchant. Vinyls are usually pigmented with aluminium, zinc chromate or red lead and are not generally suitable for brush application. The paints are one-pack and dry by loss of solvent. Despite the disadvantages vinyls are excellent paints where resistance to abrasion, chemicals, and water is required.

Preservation of woodwork

The use of wood is very limited on modern ships but it is still used for deck sheathing and other purposes on some ships. Wood decays mainly due to the action of wet and dry rot and it is generally the latter which causes problems on board ship, being caused chiefly by warmth and damp. To preserve well-seasoned wood a coating of raw linseed oil should be allowed to soak into the pores; this forms a more effective seal than boiled oil. All joints in woodwork should be filled with a suitable composition to prevent the lodgement of moisture. Dry rot may be detected by sounding with a hammer or by boring; the dullness of the sound, or the musty smell and dusty appearance of the borings, are sufficient indications of decay.

The contact of wood with steel, unless the wood can be made a complete and watertight sheathing, frequently promotes the corrosion of the steel. All woodwork connected to steel when watertightness cannot be ensured should be readily removable to facilitate inspection. To ensure watertightness, deck bolts should fit tightly in their holes, the heads must be well grommeted and bedded in white lead with close-fitting dowels over. Carelessness in caulking wood decks may lead to the formation of moisture traps which will eventually spring the bolt fastenings and cause them to leak.

The origins of leaks through wood sheathing are sometimes difficult to trace and it may be necessary to inject red lead or another composition between the surfaces to stop the leakages. After ships have been in service for a time, particularly in the tropics, the planks of sheathing often show signs of shrinkage. Local treatment by caulking may spring adjacent planks. It is better to clean out seams and butts and carefully recaulk using a sufficient number of threads of oakum or a propriety filler. If the appearance of the sheathing does not warrant extensive overhaul every third or fourth seam, or those that show signs of retaining moisture, should be hardened or recaulked from butt to butt for the length of the deck. After a long period of drying, wood decks may be treated with a coat of raw linseed oil and driers in the proportion of twenty to one.

New wood should be deeply impregnated with preservative which is applied by pressure treatment. When the vessel has been in service for some time fresh preservation can be applied by brush.

Airless spray guns

The airless spray is probably the most efficient and quickest mechanical system which the mariner can use for paint application on board ship. Paint is delivered to the gun by means of an air or electrically operated fluid pump. Atomization of the paint takes place at the gun's tungsten carbide tip solely as a result of high pressure. The fluid pressure at the gun may be as high as 3000 psi.

Various types of airless sprays are used on board ship and operators should acquaint themselves with the manufacturer's instructions on the correct usage of the equipment prior to the operation, and how to avoid risks in use. Operators should realize that different paint manufacturers' instructions should be closely adhered to. The gun should be kept perpendicular to the surface being sprayed. Avoid 'arcing' the gun as it produces an uneven coat. At the end of each stroke care should be taken to avoid swinging the gun out of its direction at right angles to the face of the work. The trigger should be released towards the end of the stroke so that the paint is feathered off and successive strokes should be overlapped. If the gun is held too close to the work surface the paint will ripple or run; if it is held too far away the paint will not be applied uniformly. If the atomizing pressure is too low the finished work has the appearance of orange peel, and if the pressure is too high the operator will be lost in a cloud of paint fog. Excessive high pressure will deliver the paint too quickly, causing sagging and paint wastage.

Safety is of paramount importance when using the airless spray and incorrect procedures or horseplay could result in a lethal accident. The gun must not be aimed at any part of the body as the paint can penetrate the skin or cause serious eye injuries. Safety procedures can be found in Chapter 14 of the 'Code'. Smoking and the use of unapproved lights should be prohibited in any space in which spraying is taking place. Suitable protective clothing such as a combination suit, gloves, cloth hood, and eye protection should be worn. A dust mask, air-fed hood, or suitable respirator must be worn. If a spray nozzle clogs, the trigger should be locked in a closed position before any attempt is made to clear the blockage. Before the nozzle is removed, or any other dismantling attempted, pressure must be relieved from the system. A gun having a reversible nozzle requires special care to ensure that hands are clear of the nozzle's orifice when a blockage is being removed by blowing through.

The spray system should be inspected regularly for defects and the pressure should not exceed the recommended working pressure of the hose. Static electricity may be generated by the pressures necessary for airless spraying and it is possible that sparking may occur between the gun and the object being sprayed; both must therefore be grounded. The hose usually contains a static electricity conductor; if not, a static wire from the gun must be attached to a suitable earth connection.

The airless spray is suitable for most paints but those containing lead, mercury, or similar toxic compounds should not be sprayed in interiors. Conventional cargo spaces should be given primer coats of yellow zinc chromate, followed by top coats of bright aluminium containing an oleoresinous low-acid base.

The tween deck spaces of a refrigeration vessel present a particular problem as the paint must be quick drying (non-oleoresinous), long lasting, high reflecting, capable of withstanding low temperatures, and should help to provide insulation by having low thermal conductivity. The paint should also be fire resistant, non-tainting, and should contain no toxic gases for safety in application. The product used is a two-pack polyurethane paint which is mixed with a chlorodifluoromethane solution (Refrige 22). The mixture reacts chemically to produce a paint with the following excellent characteristics:

ease of application and very practical with a spray gun
good durability
chemically inert
affords insulation by its expanded plastic type insulating material
bonds well to steelwork to give good corrosion protection
impervious to water transmission
fire resistant
non-toxic

A suitable wash primer should first be applied, then a sufficient number of coats of the above mixture. Heating of the area is not required if the ambient temperature is above 15°C; therefore large areas can be conveniently sprayed.

Personal protection advice for painters

The Health and Safety at Work Act 1974 places a duty on suppliers of substances for use at work to ensure as far as is reasonably practicable that those substances are safe and without risk to health when properly used. Members of the Marine Coatings Group of the Paintmakers Association of Great Britain Limited now label all drums of paint with a two-digit number. This is known as the 'Personal Protection Advice (PPA) Number' and by use of a table it informs the operator what protective equipment should be worn when applying the paint. As an example, the code number '1/6', when referred to a table, informs the operator that if he is spraying the paint in a confined space he must wear gloves, eye protection, skin protection, and an air-fed hood/mask. The PPA instructions should be strictly observed.

Shipboard paint systems

The paint system applied to any part will be dictated by the environment to which that part of the structure is exposed. Traditionally the painting of the external ship structure was divided into three regions.

1 Below the light load line where the plates are continually immersed in water.
2 Between the light and statutory load lines, i.e. the boot topping area where immersion is intermittent and much abrasion occurs.
3 The topside and superstructure which are exposed to an atmosphere laden with salt spray and are subject to damage by cargo handling and general 'wear and tear'.

However, now that high-quality paints are being used for the ship's bottom, the distinction between hull regions need not be so well defined, the boot topping area is often omitted and one system used for both the bottom and the water line regions.

Internally by far the greatest problem is the provision of suitable coatings for various liquid cargo and water ballast tanks.

Below the waterline

The ship's bottom has priming coats of corrosion-inhibiting paint applied, followed by an antifouling paint. Coatings used for steel immersed in sea water are required to resist alkaline conditions. To put it more technically, an iron alloy immersed in a sodium chloride solution, having the necessary supply of dissolved oxygen, gives rise to corrosion cells with caustic soda being produced at the cathodes. In addition, the paint should have a good electrical resistance so that the flow of corrosion currents between the steel and sea water is limited.

Such requirements make the standard non-marine structural steel primer, red lead in linseed oil, unsuitable for ship use below the waterline. Suitable corrosion-inhibiting paints for ships' bottoms are pitch or bitumen types, chlorinated rubber, coaltar epoxy resin, or vinyl resin.

The antifouling paints are applied after the corrosion-inhibiting coatings and should not come into direct contact with the steel hull, since the copper and mercury compounds present in the paint may cause corrosion. Antifouling paints consist of a matrix or vehicle in which the toxin is evenly mixed but there are two basic types: (a) only the toxin is dissolved into the sea water, (b) both the toxin and its vehicle are dissolved.

Most long-life antifouling vehicles are now formulated to maintain an unbroken surface as they release their toxic material and this presents a smoother hull surface than would otherwise be possible. Skin friction is believed to account for 80 percent of resistance to movement through water. Shipowners are increasingly aware of the importance of hull smoothness—corrosion, flaking and blistering, and fouling all cause hull roughness problems. Research has produced two completely new antifouling systems: (a) reactivatable, and (b) self-polishing paints.

The toxic properties of antifouling paints, which kill or repel the infectious stages of fouling organisms, will eventually become exhausted. A leading paint manufacturer has developed a method of using reactivatable paint by means of which vessels can be kept at sea for up to five years without fouling occurring. A high-quality anti-corrosive paint is followed by thick coats of antifouling. When the toxic level of the outer layer becomes too low, the surface is reactivated by removing the outer layer and exposing fresh antifouling. The coating is cleaned and reactivated by divers using special automatic equipment and is carried out at approximately yearly intervals. Up to three reactivations can be achieved from the initial application.

The same manufacturer produces a self-polishing antifouling which uses a specially formulated organo-tin-acrylcopolymer as a binder which slowly dissolves in sea water. The antifouling agent is thus released at a constant rate during the period between drydockings and water turbulence smooths any roughness peaks on the film surface. Self-polishing paints have a life of approximately four years.

The most serious factor leading to hull roughness seems to be the sandwich coatings which build up after a number of drydockings, a heavy layer consisting of alternate coats of primer and antifouling. One of the great advantages of reactivatable and self-polishing paints is that it is not necessary to use a sealer or primer before applying fresh antifouling, thus sandwich coatings do not build up. However, if the above systems are being applied to a ship which has been in service for a number of years it must be blast cleaned before the new system is applied.

Topsides

The modern practice requires a paint system for the hull above the waterline. These systems are often based on vinyl and alkyd resins or on polyurethane resin paints.

Superstructures

Red lead or zinc chromate based primers are commonly used with white finishing paints. These are usually oleoresinous or alkyd paints which may be based on 'non-yellowing' oils. Linseed oil based paints which yellow on exposure are generally avoided on modern ships. Zinc chromate paints should be used on aluminium superstructures.

Two modern paint systems are shown in Tables 3.2 and 3.3.

Table 3.2   Paint system for GCV—vinyl

Preparation   New ships—metal should be grit blasted to white metal finish.
Film thickness   Dry film thickness given in microns. Paint applied by airless spray.

The 1 × 25 microns barrier is a zinc-rich epoxy to protect the steel.
The 3 × 80 microns vinyl tar gives the steel further protection as it is very water resistant and it provides a good base for the antifouling coats.

Area	Primer coats	Intermediate coats	Finishing coats
Hull under water	1 × 25 Barrier	3 × 80 Vinyl tar	2 × 75 Antifouling
Hull topsides	2 × 60		1 × 40
Weather decks	2 × 60		1 × 40
Superstructures	2 × 60	1 × 40 (Optional coat)	1 × 40

A conventional system should be applied to accommodation, engine room and cargo spaces

Engine room	2 × 40 Red lead	1 × 30 Undercoat	1 × 30 High gloss
Accommodation	2 × 40 Zinc chromate in alkyd medium	1 × 30 Semi-gloss alkyd	1 × 30 Pastel shade alkyd gloss
Cargo holds	2 × 40 Zinc chromate in alkyd medium		1 × 40 Bright aluminium

Tables 3.2 and 3.3 are reproduced by permission of Jotun-Henry Clark Limited, Marine Coatings.

Table 3.3   Paint system for GCV—epoxy

Preparation   New ships—metal should be grit blasted to white metal finish.
Film thickness   Dry film thickness given in microns. Paint applied by airless spray.

Area	Primer coats	Intermediate coats	Finishing coats
Hull under water	2 × 100 1 × 50		2 × 75 Antifouling
Hull topsides	1 × 50	1 × 100 High-build undercoat	1 × 50
Weather decks	1 × 50	1 × 100 High-build undercoat	1 × 50
Superstructures	1 × 50	1 × 100 High-build undercoat	1 × 50

Table 3.3 (contd.)

Area	Primer coats	Intermediate coats	Finishing coats
Cargo holds	1 × 100		1 × 100
Ballast tanks		2 × 180 Coaltar epoxy	
Other tanks and cofferdams		2 × 100 Conventional epoxy	

A conventional system should be applied to accommodation and engine room spaces.

Further reading

BRITISH STANDARDS INSTITUTION. *Commentary on Corrosion at Bimetallic Contacts and its Alleviation* (PD 6484: 1979).

BARTON, K. *Protection against Atmospheric Corrosion* (Wiley: London, 1976).

CHAMBERLAIN, J. and TRETHEWAY, K. R. *Corrosion* (Longman: London, 1988).

FREDERICK, S. H. *Surface Preparation of Ship Plate for New Construction*, Maritime Technology Monograph No. 3, (The Royal Institution of Naval Architects: London, 1976).

MUCKEL, W. *The Design of Aluminium Alloy Ships' Structures* (Hutchinson: London, 1963).

PLUDEK, V. R. *Design and Corrosion Control* (Macmillan: London, 1977).

ROGERS, T. H. *Marine Corrosion* (Newnes: London, 1968).

SHREIR, L. L. (ed) *Corrosion* (2 vols.) (Newnes: London, 1963).

WARREN, N. *Metal Corrosion in Boats* (Stanford Maritime: London, 1980).

WEST, J. *Protection of Ship's Hulls*, Maritime Technology Monograph No 2 (The Royal Institution of Naval Architects: London, 1976).

WOODS HOLE OCEANOGRAPHIC INSTITUTION. *Marine Fouling and Its Prevention* (Woods Hole: Massachussets, 1967).

4

Ship Maintenance: Planned Maintenance

Reasons for planned maintenance

1 The most important facet of planned maintenance must be 'preventive maintenance'. This is the maintaining of the vessel and its equipment in good operating condition by the necessary continuous assessment and action. The old adage 'prevention is better than cure' is extremely important to mariners due to the isolated nature of the employment and to the lack of immediate repair and spares facilities. Good preventive maintenance means that large sums of money are not wasted on emergency spares, overtime, emergency services such as tugs, delays to the ship, and on operational incidents. Equipment, such as a component part, is replaced before the entire unit breaks down. After a certain period of time it may be cheaper to replace some components rather than constantly repair them. In general, systematic maintenance should mean fewer breakdowns and repairs.
2 Planned maintenance ensures the reliability of equipment, the equipment operates efficiently when it is required to do so and the ship's staff can rely on it in an emergency.
3 A plan ensures that the crew is working to maximum efficiency and working in most weathers and conditions.
4 The plan continually assesses the efficiency of equipment and all aspects of its maintenance.
5 A good plan ensures that no areas of the vessel or items of equipment are neglected or overlooked.
6 Well-maintained equipment presents fewer hazards to the crew.
7 Well-maintained equipment will last longer and so reduce capital outlay for the shipowner.
8 The ship will be ready to undergo surveys at short notice.
9 The ship will be up to standard should any snap inspections be conducted by appropriate authorities.

10 Planned maintenance ensures that the rate of deterioration of equipment is continually monitored, assessed, and to some extent controlled.
11 A well-documented plan ensures that new personnel are aware of the maintenance situation.
12 The use of planned maintenance means that equipment can be overhauled at convenient times. Such maintenance can be 'running', i.e. the unit is still in service, or 'shut down'. In the latter case it is important that the maintenance is carried out at an appropriate time and that the necessary spares for such work are available.
13 Planned maintenance can reduce or obviate the need for costly shore labour.

Construction of a planned maintenance schedule

Some companies operate sophisticated systems which go into elaborate and minute details of the day-to-day maintenance of the ship's equipment. One should not disparage such systems as they are of great assistance to the keen officer. However, some schemes tend to inflict an inordinate amount of paperwork on the Chief Officer and many officers feel that some schemes control the men, instead of the men controlling the scheme. Planned maintenance need not involve extensive paperwork but some basic points should be borne in mind:

1 A plan must be adaptable to various weather conditions.
2 The plan must be flexible so that changes of orders or cargoes do not upset it unduly.
3 The length of voyages, routes, and trades that the vessel is involved in must be considered.
4 The maintenance of safety equipment and emergency team training should be integrated with the overall maintenance plan.
5 The plan should be constructed so that the appropriate equipment is brought up to optimum condition for statutory and classifications surveys such as 'Safety Equipment', 'Load Line', and 'Lifting Appliances'.
6 Drydocking and repair periods should be integrated with the plan.
7 Manufacturers' advice should be complied with and all manufacturers' maintenance logs should be completed.
8 The plan should include the availability of appropriate equipment for breakdown maintenance due to unforeseen circumstances.
9 Provision must be made for spare part replacements due to wear and tear maintenance. There should also be a method for ordering spares as soon as replacement items are used.
10 The plan must be carefully thought out, well controlled, and an efficient recording system must be kept up to date.

Shipboard Operations

The plan should be broken down into various 'time phases'. Each schedule will, of course, be tailored to fit each particular ship but a schedule could be based on the following three categories:

(a) Short-term maintenance

(b) Long-term maintenance

(c) Maintenance due to operational requirements

The following is a very basic planned maintenance schedule for a general cargo vessel.

Planned maintenance schedule for a general cargo vessel

(a) Short-term maintenance

Weekly inspection and greasing (when possible)
Winches and windlass
Oil baths, if any, in winches and windlass
Wheels on steel hatch covers
Door hinges on mast houses
Ventilation system flaps and ventilators
Cleats on external weathertight doors
Anchor securing arrangements
Booby hatches to cargo holds
Sounding and air pipes
Fairleads, rollers
Derrick heels

Fortnightly inspection and greasing
Accommodation ladder and gangway
Lifeboat falls and blocks
Davit pivot points
Fire hydrants and monitors
Fire hose box hinges
Quick release gear on bridge wing lifebuoys
All lifebuoys
Liferaft securing arrangements
Securing bolts on international shore connection
Steel hatch cross joint and quick-acting cleats
Hatch gypsy drive wheels and followers
Hatch contactor panel fuses, electric cables and connections, motor heaters
All external butterfly nuts
All external electric cables and deck-lighting arrangements

128

Monthly inspection and greasing where necessary
Lifeboat falls for broken strands
CO_2 cylinders in gang release system
Fire detection system
Breathing apparatus and associated equipment
Ladders on masts and ventilation posts
Radar mast rigging
Fire gauzes
Freeing ports
Scuppers
Hatchway non-return valves
Ship side guard rails

(b) Long-term maintenance

Three-monthly inspection and/or overhaul
All cargo gear
Navigation light connections
Hold ventilation systems

Six-monthly inspection and/or overhaul
Cargo winches
Strip all mooring rollers
Fresh-water tanks
All running gear, strip blocks and derricks
Cofferdams and void spaces
Forepeak and afterpeak
Remove ventilator cowls and grease the coaming, test damper flaps and locking screws
Hold equipment such as spar ceiling, limber boards, double-bottom manholes, wells, bilges and strum boxes

Yearly
Derust and repaint derricks
End-for-end lifeboat falls
Watertight seals on hatchways
Loosen spare anchor securing bolts, lubricate all anchor parts and re-secure
Rotational cleaning and painting of storerooms, alleyways, cabins and mess rooms
Strip the windlass and aft mooring winch
Standing rigging

(c) Operational maintenance

To be carried out when necessary

Anchor cable markings

Check mooring ropes and wires before and after use

All gantlines before being used on stages

Pilot ladders and hoists, gangways, accommodation ladders, and associated equipment before and after use

Check anchors and cables stowed properly

Test fire fighting appliances before entering port

Test hand and emergency steering arrangements before approaching coasts or heavy traffic areas

Cargo securing arrangements

All cargo gear and hatch closing arrangements before and after use

Clean holds, remove all stains and touch up the ship's side. Test tween deck scuppers by pouring water down, test gas smothering pipes with compressed air, clean tween deck hatch trackways

Check the hydraulic oil in any cargo or hatch systems

Fumigate and spray holds as necessary

Regulations applicable to the maintenance of cargo-handling equipment

The use of common sense is probably the best operational and maintenance guide that mariners have but in addition all maintenance should be based on a knowledge of the appropriate statutory regulations and requirements. The appropriate regulations for cargo-handling equipment are the Merchant Shipping Hatches and Lifting Plant regulations operative 1 January 1989, which were discussed in Chapter 2. I consider it essential that all personnel involved in cargo-handling operations and maintenance should study M1347 as the Annex contains the revised Chapter 17 of the Code which set out the principles and guidance for putting the Regulations into practice. Only a few brief points are considered below.

Routine maintenance

It is clearly pointed out that the 'requirement for maintenance means that the lifting plant should be kept in good working order, in an efficient state and in good repair'. This means that the principle of systematic preventive maintenance should be adopted. There should, therefore, be very regular routine inspections of all equipment by an experienced person with the inspections being at intervals related to the character and usage of the equipment. In additon: 'Safety devices fitted to lifting appliances should be checked by the operator before work starts and at regular intervals thereafter to ensure that they are working properly'. Wire ropes should also be regularly inspected and treated with suitable lubricants. Particular care should be taken with the application of the lubricant to ensure that it penetrates the wire so that internal corrosion, as well as external corrosion, will be prevented. Wire ropes should never be permitted to dry out.

Testing of lifting plant

No lifting plant on board shall be used:
(a) after installation, or
(b) after any repair or modification which is likely to alter the safe working load (SWL), or affect the lifting plant's strength or stability, without being first tested by a competent person.

No lifting appliance shall be used unless it has been suitably tested by a competent person within the preceding five years.

Most lifting appliances will be tested by means of a static test by dynamometer or by the application of a proof load. This is the only time that the SWL can be exceeded as Chapter 17 states that 'a mass in excess of SWL should not be lifted unless':

(a) a test is required;
(b) the weight of the load is known and is the appropriate proof load;
(c) the lift is a straight lift by a single appliance;
(d) the lift is supervised by the competent person who would normally supervise a test and carry out a thorough examination;
(e) the competent person specifies in writing that the lift is appropriate in weight and other respects to act as a test of the plant, and agrees to the detailed plan for the lift; and
(f) no person is exposed to danger.

Chapter 17 also states that where proof loading is part of a test the test load applied should exceed the SWL as specified in the relevant British Standard, or in other cases by at least the following:

SWL (tonnes)	Lifting appliances	Single sheave cargo and pulley blocks	Multi-sheave cargo and pulley blocks
0–10	SWL × 1.25	SWL × 4	SWL × 2
11–20	SWL × 1.25	SWL × 4	SWL × 2
21–25	SWL + 5	SWL × 4	SWL × 2
26–50	SWL + 5	SWL × 4	SWL × 0.933 + 27
51–160	SWL × 1.1	SWL × 4	SWL × 0.933 + 27
161 +	SWL × 1.1	SWL × 4	SWL × 1.1

Such tests would normally be conducted by personnel from a repair or ship yard or from a reputable firm of ship riggers. A 'competent person' would probably be someone of managerial status from such a firm.

Examination of lifting plant

Lifting plant must be thoroughly examined by a competent person after one of the above tests. In addition, no lifting plant shall be used unless it has been thoroughly examined by a competent person at least once in every 12 month period. In the latter case a Chief Officer could be considered to be a competent person. Chapter 17 states that a thorough examination 'means a detailed examination by a competent person, supplemented by such dismantling as the competent person considers necessary, and access to or removal of hidden parts also at the discretion of the competent person in order to arrive at a reliable conclusion as to the safety of the plant examined'.

A competent person is defined in Chapter 17 as someone over 18 and of 'such practical and theoretical knowledge and actual experience of the type of machinery or plant which he has to examine as will enable him to detect defects or weaknesses which it is the purpose of the examination to discover and to assess their importance in relation to the strength, stability and functions of the machinery or plant'.

Certificates and Reports

The master shall ensure that a certificate or report shall be supplied within 28 days following any test or examination and shall be kept in a safe place on board ship for a period of at least 2 years from receipt of the certificate or report of the next following test or examination. Certificates are thus required to be written within 28 days. Certificates or reports should be kept readily available and should be available to any dock worker or shore employer using the ship's plant.

M1347 contains examples of certificates, which should be in the format recommended in ILO Convention No. 152. For example, the certificate of test and thorough examination of lifting appliances should contain the following information:

    Identity of national authority or competent organization
    Certificate number
    Name of ship
    Official number
    Call sign
    Port of registry
    Name of owner
    Situation and description of lifting appliances which have been tested and

thoroughly examined
Angle to the horizontal or radius at which test load applied
SWL at the angle or radius
Test load (tonnes)
Name and address of the firm or competent person who witnessed testing and carried out thorough examination
Signature and declaration of competent person
Date
Place

Reports of the '12 monthly' thorough examinations should be written in a 'Register of ship's lifting appliances and cargo handling gear' in a format in accordance with ILO Convention No. 152. Part 1 of the Register contains entries with regard to 'Thorough examination of lifting appliances and loose gear'. Part 1 contains 5 columns:

1 Situation and description of lifting appliances and loose gear which have been thoroughly examined.
2 Certificate numbers.
3 Examination performed, e.g. Initial
                          12 monthly
                          Five yearly
                          Repair/damage
                          Others
4 Declaration by competent person, signed and dated.
5 Remarks (to be signed and dated).

Part II should contain details of regular inspections of loose gear.
    In addition to any of the foregoing, any competent person discovering a defect affecting the safety of plant should ensure that a suitable person is informed of the defect so that the defect can be remedied.

Rigging plans (ships' derricks)

Rigging plans shall be available and contain the following information:
(a) position and size of deck eye-plates
(b) position of inboard and outboard booms
(c) maximum headroom (i.e. permissible height of cargo hook above hatch coaming)
(d) maximum angle between runners
(e) position, size and SWL of blocks
(f) length, size and SWL of runners, topping lifts, guys and preventers
(g) SWL of shackles
(h) position of derricks producing maximum forces
(i) optimum position for guys and preventers to resist such maximum forces

(j) combined load diagrams showing forces for a load of 1 tonne or the SWL
(k) guidance on the maintenance of the derrick rig

Maintenance schedule for cargo-handling equipment

1 All grease nipples on winches, blocks, derrick heels, crane 'turntables', and similar equipment should be attended to weekly.
2 Inspection of ancillary equipment such as chains, rings, hooks, swivels, blocks, and shackles every 3 months.
3 A thorough overhaul of the above equipment every 6 months. The derricks or derrick cranes should be stripped of all the ancillary equipment and the gear taken apart, examined, greased, and re-assembled.
   All grease nipples should be extracted and examined.
   All items of equipment should be examined to ensure that the SWL and identifying number are legible.
   The location and identifying number of each individual item should be checked against the rigging plan location and number of that item. If any piece of equipment is replaced the rigging plan must be amended accordingly. It is essential that the actual position of all the cargo handling equipment corresponds with the position shown on the rigging plan. The file containing the certificates should be examined to ensure that each item does have an appropriate certificate.
   Winches should also be overhauled every 6 months with the co-operation of the engineer officers.
4 The lifting appliances, e.g. derricks, should be de-rusted, overhauled, and painted every 12 months. Particular attention should be paid to the goose neck swivel and it may be necessary to lift the derrick so that the goose neck can be overhauled.
5 All equipment, including wire runners and winches, should be inspected before, during and after use. If any item appears suspect it should be replaced immediately by equipment which has an appropriate certificate. The rigging plan should be amended immediately.

Care of cargo blocks

Frequently check the swivel head for free movement by hand; grease the shank and bearing. Examine the side plates for distortion or buckling; a runner could become caught between a sheave and a distorted side plate, thus causing a serious accident. Sheaves should turn freely when rotated by hand

and they should be examined for cracks and bush wear. The grooves of the sheaves should be checked for wear which will quickly ruin a new runner. Axle pins should be secure and unable to work adrift; the thread in the pin should be checked for damage. If possible avoid painting blocks as this practice clogs grease nipples and reservoirs, covers statutory markings, and hides defects. The surfaces of blocks should be oiled frequently. Self-lubricating blocks should have the reservoirs cleaned out and refilled with a suitable lubricant. Conventional blocks should be lubricated daily when in use. Check all split pins and inspect the distance pieces.

Overhauling the derrick heel goose neck

If possible this operation should be carried out when the vessel is at anchor as the derrick must be lifted in order to inspect part of the goose neck. If conducted on passage the weather conditions should be ideal, the derrick should be well secured when it is unshipped, and due regard should be had to unforeseen athwartships movement of the vessel. Before starting the job a temporary secure crutch for the derrick heel should be made so that the derrick is not left suspended on the lifting tackle. Inexperienced personnel should be instructed on their role in the operation and all applicable safety precautions should be taken.

1 Securely lash the derrick head in its crutch.
2 Remove and overhaul the derrick heel block.
3 Secure a purchase of appropriate SWL to a suitable position on the mast or samson post and to the derrick. A direct lift can often be obtained over the derrick heel by unshipping the derrick topping lift block and securing the purchase by a strap to the heel of the derrick.
4 If the goose neck securing arrangements are similar to those shown in Figure 4.1, the split pins should be withdrawn and paint and corrosion removed from around the bolts. On some ships the bolts nuts may be secured by an additional locking or ring nut.
5 Lubricate and remove the vertical and horizontal pivot bolt nuts.
6 Heave tight, preferably by hand, on the lifting purchase and take the weight of the derrick.
7 Lubricate, free, and remove the pivot bolts. A gentle tapping with a hammer may be necessary to dislodge the bolts.
8 Unship the derrick heel and secure it in the temporary crutch.
9 Clean all surfaces thoroughly and check all parts for signs of wear or hair cracks. Particular attention should be paid to the bolts.
10 Thoroughly lubricate all areas and re-assemble the goose neck area to its operational condition.

Non-patent heavy lift derricks

Due to the weight of the equipment of jumbo derricks it is usually not practicable to carry out an overhaul at intervals of less than 12 months. It may be necessary to use shore facilities to carry out an overhaul and occasionally unforeseen accidents occur. In the event of a heavy lift derrick being dropped the following procedure should be observed:

1 The derrick should be unshipped by means of a shore crane and removed for repair and examination.
2 The accident should be investigated by the Safety Officer and if it was caused by component failure, then that item should be replaced and sent ashore for examination. All other items of equipment should be inspected, overhauled, and replaced if damaged.

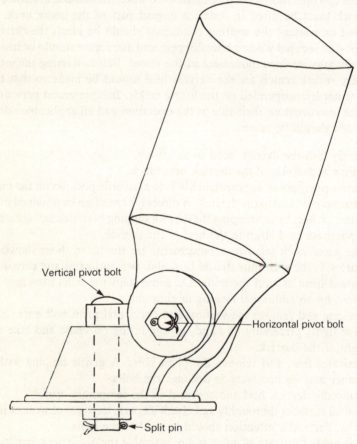

Figure 4.1   Derrick heel goose neck

3 When the derrick is repaired it must be re-rigged and tested as stated in the regulations. The test should be conducted by an approved shore establishment such as a shipyard or a company which carries out derrick work.

When the test and thorough examination are completed the standard certificate entitled 'Certificate of Test and Thorough Examination of Lifting Appliances' will be issued. The certificate should have four columns which give the following information:

(a) Equipment tested, position on the ship, identification marks of the equipment
(b) Angle to the horizontal in degrees during the test
(c) Proof load applied
(d) SWL at the angle shown

The certificate is signed by a responsible representative of the firm which carried out the test. The name, address, position, and qualification of the representative are shown on the certificate, also a statement by that person corroborating the fact that the test was carried out and that a thorough examination afterwards showed that the equipment had withstood the proof load. A 'Certificate of Test and Thorough Inspection of Loose Gear' will probably also be required.

Safety precautions when overhauling cargo appliances

Chapters 15 and 17 of the 'Code' should be consulted.

Working aloft

1 Only experienced seamen should be sent aloft. Seamen with less than 12 months' sea time or who are under 18 years of age should only be sent aloft under adequate supervision.
2 Safety harnesses must be worn.
3 Shut off power to the whistle, post notices in the radio room, and instruct the radio officer not to transmit. Request the engineers to reduce steam and smoke emissions.
4 If work is taking place near radar scanners, the radars should be isolated and warning notices put on the sets.
5 Before the men proceed aloft all supporting equipment, such as bosun's chairs and gantlines, should be inspected and load tests applied.
6 If possible, safety nets should be rigged.
7 Tools should be sent up in suitable containers.
8 The containers should be secured aloft so that tools not in use can be stowed in the containers. Tools should not be left unsecured on areas such as mast tables.

9 Suitable clothing should be worn, it must not impede movement but should not be loose. Non-slip protective shoes should be worn and a safety helmet.

10 If possible do not raise men in bosun's chairs. If it is necessary to do so they must be raised by hand. Under no circumstances should men be raised by power.

11 Working on blocks invariably means greasy hands; care must be taken when handling tools. Equipment belts might be of use to stow frequently used tools.

12 All shackles which do not have locking pins should be seized.

13 Blocks should be sent down separately on individual gantlines.

Working on deck

1 Protective clothing must be worn.

2 Derrick blocks are heavy. They should be lifted correctly with sufficient men to do the job properly.

3 Do not stand underneath men working aloft, especially when blocks are being sent down.

4 Use proper supports if working a short distance above the deck. A much abused forty gallon oil drum is not a proper support.

5 Portable ladders should be lashed before being used.

6 New wire should not be used directly from a coil. Flake 'up and down' on the deck before use.

7 Frequently remove oil and grease marks from the deck.

When work is completed all safety devices, such as split pins, should be inspected.

Care and maintenance of mechanical steel hatch covers

Weather conditions and operational problems will probably ensure that a strict adherence to the following schedule is impossible; it should be accepted only as a guide. The schedule directly applies to the single-pull hatch cover type and should be adapted for other systems such as hydraulic folding covers and piggy-back covers.

Weekly

Clean and grease the eccentric wheels.

Fortnightly

Clean and lubricate all quick-action cleats, cross-joint wedges, gypsy drive wheels and followers, and the balancing rollers. Request the duty engineer to isolate the contactor panels and check the fuses, cables and connections, and the heaters. All other electrical maintenance should be carried out by an electrical officer.

Monthly

Check the hatch coaming non-return valves by pouring water into the drainage holes in the coaming top. Inspect the drainage channels between the sections to ensure that they are not clogged with dirt. Hose test the covers with a water pressure of approximately 50 psi and inspect for leaks. Check all drainage channels and coaming trackways for signs of corrosion.

Six-monthly

Remove all corrosion from the external and internal surfaces of the covers. Look for signs of rusting underneath the hatch cover neoprene watertight seal.

Yearly

Check the seal for resilience and signs of perishing. When the hatch covers are secured the seal makes firm contact with steel compression bars welded on top of the coaming. Check the bars for signs of damage, apply chalk to the bars, close the covers and secure them. Open the hatch and check that the seal has been marked by the chalk; the seal will have to be renewed in any sections which are unmarked.

Operational maintenance and safety precautions

If the hatch has motorized panels the towing chains may need adjusting after a period in service. Before closing the hatch ensure that the trackways are clear of debris, that all cleats are in the proper stowage position, and that the panel safety securing chains are released. Before opening the hatch, free all side securing cleats, remove all cross-panel wedges on top of the hatch, turn the eccentric wheels and check the locking pins, and remove the safety locking pins. When the covers are moving all personnel should keep well clear and never stand on top of the panels. A restraining wire in good condition should be used to prevent the panels 'running away'; modern motorized panels have a braking system to prevent such an occurrence. Follow all the manufacturer's instructions regarding safety and maintenance.

Care and maintenance of pilot ladders and hoists

Before including the pilot ladder and hoist in a maintenance schedule an officer should check that both comply with the following regulations:

> SOLAS 1974, Chapter V, Safety of Navigation, Regulation 17; The Merchant Shipping (Pilot Ladders and Hoists) Regulations 1987

In particular the following points should be considered when making an initial inspection of the pilot ladders:

1 The ladder must be in one continuous length. The practice of having a short ladder for the loaded passage, which was then shackled to another length to make a long ladder for the light ship passage, is not permitted.
2 The steps must be made of hardwood or other material of equivalent strength and the four lowest steps may be of rubber or similar material.
3 The steps must be in one piece and if made of wood must be knot free. They must have a non-slip surface and the dimensions should be not less than 480 mm long, 115 mm wide, and 25 mm in depth; non-slip grooving is not included in these dimensions.
4 All the steps should be the same distance apart, the equal spacing should not be less than 300 mm and not more than 380 mm, and they must be secured in such a manner as to remain horizontal.
5 Check the ladder for replacement steps as no ladder is permitted to have more than two replacement steps which are secured in any way that differs from the original method. If the replacement steps are secured to the side ropes by grooves, such grooves must be in the long side of the steps.
6 Spreaders, which should be of the same material as the rest of the ladder and between 1.8 and 2 metres long, should be at such intervals as to prevent the ladder from twisting. However, the fifth step from the bottom should be a spreader and the intervals between any spreaders must not exceed nine steps.
7 The side ropes must consist of two continuous manila ropes on each side with a diameter of not less than 18 mm.

Any ladder which does not comply with SOLAS standards must be replaced as soon as possible. The regulations also contain certain safety provisions regarding the rigging, use, and access from pilot ladders and the crew should be instructed on the following procedures:

1 Two man ropes of not less than 20 mm in diameter should be provided securely rigged to the ship.
2 A safety line with a harness, a lifebuoy with a light, and a heaving line should be kept near at hand ready for use. Do not attach the heaving line to the lifebuoy.

3 The ladder should be rigged:
   (a) clear of any discharges
   (b) clear of the finer lines of the ship and as near midships as possible, but not under any overhanging parts of the ship's hull structure
   (c) so that each step is firmly against the ship's side

4 The person using the ladder should not climb less than 1.5 m and not more than 9 m. If the distance from the sea level to the point of access to the ship is more than 9 m, an accommodation ladder must be rigged so that the pilot can transfer to the accommodation ladder. The pilot ladder's upper end must extend at least 2 m above the accommodation ladder's lower platform; the latter should be rigged as near to the mid-length of the ship as is practicable, it should lead aft, the lower end should rest firmly against the ship's side, and the falls should be bowsed in.

5 The area should be well lighted.

6 The deck area should be free of grease.

7 A boat rope should be kept at hand.

8 The ladder should not be secured to the side rails but to cleats or eye pads on deck.

9 Safe access must be provided from the head of the pilot or accommodation ladder to the deck. If access is by means of a gateway in the rails or bulwark, adequate handholds shall be provided. If access is by means of a bulwark ladder, it must be securely attached to the bulwark rail or landing platform and to the deck. Two handhold stanchions, which are not part of the bulwark ladder, must be rigidly secured to the ship's structure. Each stanchion must be at least 40 mm in diameter, and be secured to the ship at its base and at a higher point; it must extend at least 1.2 m above the top of the bulwark, and the stanchions should be spaced between 0.7 and 0.8 m apart.

10 The rigging of ladders, and the actual embarkation or disembarkation, must be supervised by a responsible officer.

11 It must be possible for the ladder to be rigged either side of the ship. Thus if the ship is carrying deck cargo, means must be provided to enable the pilot to board on either side. Each pilot ladder must be kept clean and in good order and must only be used in pilot operations, or by officials and other persons while a ship is arriving at or leaving a port.

Pilot hoists are sometimes fitted on vessels which have a very large freeboard when in ballast conditions. The hoists are built to stringent specifications and must be of a type approved by the Secretary of State. A hoist and its ancillary equipment must be of such a design and construction as to ensure that it can be used in a safe manner. A safe access from the hoist to the ship must be provided. The hoist must only be used in pilot operations, or by officials and other persons while a ship is arriving at or leaving a port.

A hoist must be clearly marked with the maximum complement it is permitted to carry, it must be located clear of the finer lines and as close as practicable to midships but not underneath any overhanging parts of the ship's hull structure, and the operator at the control point must be able to see the whole operation. A copy of the manufacturer's maintenance manual, which contains a maintenance log, must be kept on board and an officer must enter a record of maintenance and repairs in the manual. The hoist must be well maintained and kept in good order. The construction of the hoist consists of three main parts:

(a) a mechanically powered winch
(b) two separate falls
(c) a ladder consisting of a rigid upper part on which a person stands and a flexible lower part of a short length of pilot ladder which enables a person to board from or disembark to a launch

(a) The source of power can be electric, hydraulic, or pneumatic. The source of power must not cause a hazard to a ship carrying flammable cargo and if pneumatic must be provided with an exclusive air supply. The winch must include a brake which can support the working load in the event of a power failure and hand gear capable of lowering a person 'stranded' in the same circumstances. Engagement of a manual crank handle must automatically cut off the power supply and the hoist must have a safety limit (cut-off) device and an emergency stop. In addition, the person being carried must be able to operate an emergency stop switch. The speed of the hoist should be between 15 and 30 m per minute and the falls must wind evenly on to the winch drums. The controls must be clearly marked and the hoist securely attached to the ship's structure (not to the side rails). The means of access must be by a platform guarded by handrails, and any electrical appliance associated with the ladder section must not be operated at a voltage exceeding 25 volts.

(b) The falls must be of flexible steel wire rope of adequate strength and which is resistant to the corrosion-inducing environment in which ships operate. They should be securely attached to the winch drums and to the ladder and must be of such a distance apart as to reduce the possibilities of twisting. The falls must be long enough to do the job in all service conditions and still leave three turns on the winch drum.

(c) The ladder section must have a rigid part at least 2.5 m long with sufficient non-skid steps to provide safe and easy access and with safe handholds. A spreader with rollers is fitted at the lower end to enable the ladder to roll freely on the ship's side and an effective guard ring is fitted to provide physical support for the person on the ladder. Adequate means for communication and the emergency stop switch must be provided on the rigid part. The flexible lower part must be eight steps long and constructed to the same specifications as for a normal pilot ladder.

Every new hoist must be subjected to an overload test of 2.2 times the working load and after installation on board an operating test of 10 percent overload must be carried out. The Chief Officer should bear in mind that a hoist is examined under working conditions at each survey for the renewal of the 'Safety Equipment Certificate' and should carry out maintenance accordingly. He should also note that regular test rigging and inspection, which should include a load test to 150 kg, must be carried out every six months and an entry to that effect made in the ship's official log book.

As with the pilot ladder, certain operational procedures regarding the hoist should be brought to the crew's attention:

1 Any crew member involved in any operational aspect of the hoist must be given relevant instructions.
2 The hoist should be rigged well in advance of the time at which it is to be required and tested before use.
3 A competent officer must supervise the rigging, testing, and operation of the hoist.
4 The operational area must be adequately illuminated.
5 The pilot ladder should also be rigged and ready for use.
6 A safety line and harness, a lifebuoy with light, and a heaving line must be kept at hand ready for use.
7 The position on the ship's side where the hoist is lowered should be marked.
8 An adequately protected stowage position must be provided for the hoist.

Mariners should bear in mind that a ship may be detained if the pilot ladders or hoist do not conform with the regulations. The owner and master can be guilty of an offence if the equipment is not provided or up to standard and an officer can be guilty of an offence if he does not supervise or maintain equipment properly.

M898, 'Pilot Ladders and Mechanical Pilot Hoists', should be studied in full. The International Maritime Pilots' Association requires that all vessels brought into service after 1 July 1980 provide a pilot access point on each side of the vessel.

Problem of 'frozen' fairleads

Fairlead rollers which become rusted and will not turn are usually the consequence of neglect but the problem can occur on well-maintained ships when weather and operational conditions do not permit the weekly greasing programme to be carried out. Rollers which are seized cause excessive damage to mooring ropes and must be overhauled as soon as possible. The basic design varies from ship to ship but most roller fairleads are similar to the type shown in Figure 4.2. However, some fairleads have upper and lower spindle bolts instead of the one 'through' spindle in the diagram.

The grease reservoir covers should be removed, the old grease taken out, and the reservoirs filled with a release and penetrating oil. All surfaces in contact with another surface should be given a liberal application of oil and left to soak overnight. The following morning use a gantline size rope to put a round turn on the roller, lead the rope to a winch, and heave gently. This is usually sufficient to free all but the most obstinate roller.

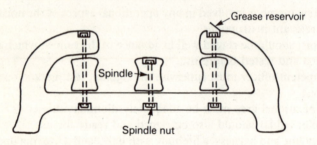

Figure 4.2   Roller fairleads

If this is unsuccessful, the roller must be efficiently secured and the spindle withdrawn. Care should be taken when loosening the nuts; use plenty of release oil and only apply gentle persuasion such as tapping with a hammer. On some vessels it may be possible to heat the nuts in order to expand them and to loosen the rust. The engineers will probably be able to lend specialized tools for withdrawing the spindles. On removing the roller it will be observed that the rust will be concentrated on the horizontal surfaces and on the spindle. Remove all corrosion and check that the grease tracks are undamaged. It may be necessary to renew the spindle nuts and washers and possibly even the spindle itself. The horizontal surfaces should be protected by several coats of primer followed by two topcoats and heavy duty grease should be applied to the spindle. Grease nipples should be unscrewed and cleaned. When the roller is reassembled check that grease can be forced through the grease nipples on the appropriate surfaces. Resume the weekly greasing schedule.

Planning for heavy weather damage

The emergency team or damage control party should be trained to deal with damage caused by heavy weather as on most ships this is the most frequent type of damage to be encountered. A useful exercise is to inform the party that heavy seas have carried away some of the ship's side rails and have also fractured a hold ventilator at deck level. A variation of the following procedure could be used to assign duties to crew personnel:

1 Temporary guard rails should be constructed by using suitable size wire rope with supports such as spare gangway stanchions, dunnage or timber (8 × 8 cm is a useful size), or angle bars. It may be possible to weld sockets to the deck in order to provide support for the timber or the angle bars could be welded directly to the deck. At least two guard rails, the upper at a height of 1 m, should be set up as tight as possible. The area in the vicinity of the damaged rails should be fenced off and warning notices posted.
2 Sound the bilges in the immediate area and carry out a damage check.
3 Remove any loose portions of the ventilator shaft but leave any sound parts to help support a cement box.
4 A wooden frame should be built around the base of the shaft or around the hole if the complete ventilator has been washed overboard. The frame should be constructed of short lengths of wood approximately 3 × 10 cm to a height of 30 cm and the corners of the frame may be supported by the use of angle bars. The planks should be joined by vertical flat bars and once again, if possible, angle bars can be welded to the deck to support the frame. If necessary the frame may be tommed off to hatch coamings or mast houses by lengths of 8 × 8 cm (3 × 3 in) timber.
5 It may be necessary to construct a temporary base for the cement by placing wood underneath the hole in order to support the wet cement during the drying period. If possible this should be done from the tween deck and toms can be left in position to support the box.
6 Drill the ends of any cracks which may appear in adjacent deck plating to prevent any extensions occurring. The holes and cracks should be filled with plastic steel compounds.
7 Dry the box and surrounding area. Lime and soda may be used as drying agents.
8 Mix cement in the proportion of three parts of fine sand to one of cement and fill the box. If the weather is inclement, it may be necessary to rig temporary protection to enable the sand to dry out.
9 Inspect the damage area at frequent intervals.

If the ship has the facilities a doubling plate could be welded or fitted over the ventilator hole. If welding equipment is not available it would be necessary to remove the tween deck ventilator shaft. One doubling plate with pre-drilled bolt holes must be placed over the deck opening and a similar plate held against the underdeck frames. Bolts of suitable length can then be inserted through both plates and secured. The deck doubling plate should be made watertight by use of a suitable compound.

The above operations can only be carried out when the ship is hove to or when the weather conditions are suitable.

At the next port the damage should be repaired by a reputable company. As the watertight integrity of the ship has been affected and also the conditions of freeboard assignment, the damage and the repairs must be inspected by a surveyor from the assigning authority.

Repair and drydock lists

An important aspect of planned maintenance is continuous repair assessment and the preparation of a drydock repair list. The ship's staff should systematically compile comprehensive and accurate details of all items requiring overhaul and repair. The description of the nature of the repair should be precise. If a Chief Officer on a tanker stipulates 'Repair number 3 main cargo pump', the pump will be stripped, the parts will be machined, overhauled, and repaired if necessary, and the pump will be re-assembled and restored to an 'as new' condition. If the only problem with the pump had been a leaking gland the Chief Officer will have caused the company considerable expense and will deservedly receive an admonition from the Superintendent. The specification for a repair item should include accurate dimensions, descriptive details, owner's spares available, information regarding parts which the ship's staff suspect may need renewal, information pertaining to the location of the item, and any access problems or possible use of staging. Sketches should be made when appropriate and it is often helpful to attach photographs from the ship's 'instant print' camera.

Some companies require the repair list to be presented under separate headings such as:
(a) standard items
(b) repair items
(c) modification items

(a) Standard items

1 Hull cleaning, surface preparation, painting.
2 Inspection and overhaul of anchors and cables, including ranging and marking.
3 Inspection, cleaning and painting of cable lockers.
4 Plugs to be taken from all bottom and peak tanks. (The plugs should be

labelled and retained by the Chief Officer and replaced before the dock is flooded.)

5 All sea valves and sea chests to be inspected, overhauled and painted.
6 Inspection and overhaul of rudder and propeller.
7 Inspection, overhaul and load test of all lifting appliances.
8 All tanks, holds, compartments, and their closing appliances to be inspected and overhauled.
9 All anodes to be inspected, the location and weight or size to be ascertained.
10 Survey of ship's bottom (known as 'sighting the bottom') to be conducted.

(b) Repair items

1 Renewal of piping.
2 Cargo-handling equipment.
3 Hatch-closing arrangements.
4 Bulkhead leaks.
5 Hull structure damage.
6 Replacement of ship's side rails.
7 Instrumentation and control equipment refurbishing.
8 Electric cables.
9 Heavy weather damage.
10 Overhaul of fire-fighting and life-saving appliances.

(c) Modification items

1 Fire-fighting systems such as foam or carbon dioxide.
2 Fire detection system.
3 New piping and structural arrangements, e.g. 'segregated ballast tank' system.
4 Inert gas systems.
5 Life-saving appliances arrangements.
6 Conversions or restructuring in order to comply with any new mandatory equipment requirements.

Organization of the ship's staff in repair yards

Most shipyards will provide the following basic facilities:
    access
    water for the ship's fire main
    electricity

compressed air
fire watchmen
steam
daily garbage removal
means of keeping the ship's refrigeration unit in operation
telephone
protective coverings for alleyways and furniture
toilet facilities (within a reasonable distance from the ship)
means of providing heat in accommodation areas
tugs and riggers when moving the vessel
facilities for cleaning and removing residues of slop tanks
liability and insurance cover

If the vessel has a full crew on stand-by the above points should be considered when the crew utilization plan is being prepared. One of the Chief Officer's (or Safety Officer's main preoccupations should be safety. In some shipyards the amount of maintenance work which the ship's staff are permitted to carry out is limited; therefore, full use should be made of the availability of crew so as to ensure a high standard of safety. It is essential that an officer liaises with the shipyard staff to check what maintenance may be carried out by the ship. However, the following routine work and precautions should be carried out:

1 Personal safety The crew should be instructed on general safety requirements and the procedures to be taken in the event of an accident or fire breaking out. Protective clothing should be worn by all personnel; particular attention should be paid to the wearing of safety helmets and shoes.

2 Access Although the dock company will normally provide the access facilities, a ship's watchman must be in attendance at all times. An officer must check that the access arrangements are suitable and he should apply the ship's access equipment criteria, which are discussed in Chapter 2, to any shore gangways. The ship's officers must consider themselves to be at all times responsible for the ship's access arrangements. The ship's watchman should be instructed on his duties and responsibilities. Notices should be posted at the gangway so that personnel can readily ascertain the methods for summoning fire and ambulance services. The location of the nearest telephone should be stated and a list of relevant telephone numbers should also be given.

3 Hot work Before any hot work is commenced a reliable person must inspect the immediate area for flammable material. Adjacent areas must also be inspected, especially if a bulkhead is to be welded, and any materials likely to constitute a fire hazard must be removed.

4 Fire precautions The vessel's own pumps will be inoperable in drydock and the fire patrol should check that the fire main is kept pressurized by water supplied by the shore, using the international fire connection if necessary. No fire-fighting appliances, such as portable extinguishers, should be sent ashore until replacements are supplied. M825, 'Precautions to be taken to prevent the accidental release of CO_2 fixed fire smothering systems', should be studied. This notice states that CO_2 systems have been accidentally activated during repair periods with the consequent endangering of lives and lists precautions which should be observed to prevent such occurrences.

It is generally accepted that all fixed systems should be locked to prevent accidental release. The keys should be held by the duty officer. The fire patrol must be aware of the duty officer's location and should report to him at suitable intervals.

M825 also emphasizes that shipowners should 'be aware that responsibility for fire protection and initial fire-fighting measures' generally remains with the vessel. If the responsibility has been delegated to the repairers there must be a written agreement to that effect and the terms of agreement should be made known to all the parties concerned.

The fire patrol must be instructed on the alarm procedures and initial measures to be taken in the event of a fire being discovered. The patrol should realize the importance of carrying out an inspection immediately after repair staff have left the vessel at the end of a shift. Fires have broken out in areas recently vacated by workmen.

5 Tank inspections Cargo and bunker tanks are particularly dangerous areas due to the possibility of a flammable or explosive atmosphere being present. All tanks which have been sealed for some time may be deficient in oxygen or contain an excess of hydrogen. Before work commences the tanks will have been cleaned and ventilated and a chemist will have issued certificates permitting 'hot' or 'cold' work in such compartments. In addition, a chemist should inspect the cargo compartments on tankers every morning before work commences and also in the evening to ensure that the tanks are in a suitable condition for either hot or cold work. A card, which gives details of the twice daily inspection, is usually attached to the tank entrance lid. An officer should be detailed to check that such inspections have been properly carried out.

6 Deck watch There should be sufficient crew members on board at all times to ensure that a 'deck watch' is maintained. At night the watch would assist in dealing with any emergencies and during normal working hours can load stores, prepare the vessel for surveys, and assist during the actual surveys. Experienced Chief Officers know that a myriad of small jobs frequently arise during repair periods and that it is necessary to have a petty officer and some men available to deal with such problems.

7 *Keeping a clean ship* It is extremely difficult to keep a ship clean and tidy during a repair period. However, a confused muddle of gear lying about the decks is a safety hazard and efforts must be made to keep the decks clear. Equipment can often conceal dangers; deck clutter is a particular danger on tankers when many tank openings are left uncovered as an aid to ventilation. A crew member could be instructed to make a twice-daily check that all openings into which a person could fall are fenced to a height of 910 mm. Other crew members may be required to assist with the daily garbage removal and with cleaning up oil spills on deck.

All movable items of ship's equipment should be locked in the storerooms and one man should be put in charge of stores. All items of ship's gear which are issued to drydock staff to enable them to carry out repair work should be signed for and a careful note made of the issuing details.

All the ship's toilets should be locked as unauthorized use is a health hazard and could be detrimental to the relationship with the repair staff. If possible, shower areas and similar facilities should be locked to prevent them being used as urinals.

8 *Ship's lighting* The watchman, or other crew members, should check that the vessel is adequately lighted and that the numerous cables lying about the various decks are not a safety hazard.

9 *Portable heaters* are often used in the accommodation and the fire patrol should check frequently the areas in which heaters are being used.

10 *Safety checklist* Some yards have a ship/shore 'Safety Checklist' which is somewhat similar to the list used in oil terminals. An officer should carry out the first check and a responsible person should carry out subsequent checks at suitable intervals.

The above duties should be discussed at the daily staff meeting and should be incorporated into the crew's duties for that day.

Delegation of repair period duties

The Chief Officer would be foolish if he attempted to carry out many of the routine duties himself; indeed, it would be impossible for him to do so. Before commencing the repair period the Chief Officer should have a staff meeting during which he can instruct the other deck officers, the petty officers, and the cadets on their responsibilities. The Chief Petty Officer should give advice on which crew members should be watchmen, should be in charge of stores, take part in fire patrols, and the other routine tasks. The Safety Officer should have a cadet to assist him in his extensive duties and safety should be his main preoccupation. All the officers should have their general duties outlined:

ensuring the safety and security of the vessel
checking the appropriate numbers of men involved in the various jobs
    and the time spent on those jobs
inspecting completed work before 'boxing up' commences
witnessing tests
assisting with and conducting surveys
checking on materials consumed on specific jobs
checking on the location of ship's tools and the use of ship's spares

The officers should have copies of the repair lists and should familiarize themselves with any particular job which they are monitoring. Repair items should be identified with the working specification numbers and officers should be conversant with the nature of any repair work being carried out. The Radio and Electronics Officer will probably be able to assist with the monitoring of any work being conducted on the electronic navigation equipment. A deck officer is usually involved with supervising the application of any protective or special coatings. He should ensure that the surface has been properly prepared, that the paint has been adequately mixed, and should note the spraying pattern, the film thickness, and the amount of paint used.

When making out the duty roster the Chief Officer should, if possible, give his personnel adequate shore leave. During the repair period the Chief Officer should have a daily conference with the repair staff and with his own staff. The use of portable radios aids the smooth running of the repair operation.

Preparing a vessel for a repair period after discharging a sulphur cargo

A vessel cannot enter a drydock until all the sulphur residues have been removed from the hold. The main hazard during cleaning is the risk of dust explosion. Fine grain sulphur presents an additional hazard in that the explosive mixture can be ignited by static electricity. The permit to work system must be used throughout the operation of hold cleaning. To prevent the hold atmosphere becoming dust laden the initial cleaning should be a wash down with hoses producing a good pressure. Attention should be paid to areas which are inaccessible or difficult to reach; box beams are particularly difficult to clean. The mixture of water and sulphur can cause corrosion problems and the operation must be carried out as quickly as possible. If the residues are being pumped overboard in port, permission should be obtained from the appropriate authority before pumping commences. The holds must be adequately ventilated during the cleaning process.

After hosing, it will be necessary to sweep the holds and to remove residues from the bilges. The hosing down operation may have to be repeated until a suitable standard of cleanliness is obtained. The ship will be inspected by a drydock representative before permission is given to enter the repair area. All pumps and lines should be thoroughly cleaned at the end of the operation and personnel must wear protective clothing and face masks during the cleaning process. As with all cargoes of this nature the 'IMDG Code' should be consulted. The Code recommends that fine grained sulphur should not be carried in bulk.

Inspecting the forepeak tank on new building or before leaving drydock

The following operation should be conducted by several people and 'Dangerous Spaces' procedures should be observed:

1 Check that no rungs are missing from any ladders.
2 As many welds as possible should be checked. It is practically impossible to check all the welds in the forepeak of a large vessel but weld failure can occur in peak tanks and must be guarded against.
3 Inspect any protective coating and ensure that areas which are difficult to reach have been adequately covered.
4 If sacrificial anodes have been fitted, check that the anode positions agree with the plans and that the anodes are secure.
5 Ensure that the sounding pipe is correctly located and that the striker plate has been fitted. Have a sounding rod lowered through the pipe and view it touching the striker plate.
6 Check that the drain is correctly located and in the position indicated on the plan. Make sure that the screw drain plug actually moves and that it has not been tack welded in position.
7 Check that the air pipes and filling pipes have been fitted with appropriate plugs.
8 Make sure that all loose equipment and shipyard rubbish has been removed.
9 The pumping arrangements should be given a thorough inspection.

Check the pump for any loose bolts and defective workmanship. Inspect valve suctions, extended spindles, and strum boxes. Look for inappropriate or poorly insulated bimetallic connections. When satsified with the standard of workmanship, give the pump a dry run by slowly starting it and feeling for air movement in the suction. The tank should then be filled to a depth of approximately 1 m and pumped dry. The pump should be checked for operational deficiencies and the pipe connections checked for leaks.

10 The Chief Officer should be present with the Surveyor at the 'Tank Test'. Extension pieces are fitted to the filling pipes and the tank slowly filled until a head of 8 feet or 2.45 m above the top of the tank is obtained. Great care should be taken to avoid over-pressurizing the tank. Bulkheads, cofferdams, watertight seals on the manhole covers, and all areas adjacent to the forepeak should be checked for leaks. The water in the tank should then be dropped to the operational level. This test may be carried out afloat so as to reduce the stress on the ship's structure.

A note for potential candidates for DoT Certificate of Proficiency/HND Nautical Science

Valuable maintenance experience can be obtained while at sea by noting the planned maintenance schedule of the vessel on which you are serving and by constructing your own schedule.

A copy of the vessel's overall paint system should be made together with any maintenance problems associated with that system.

The construction of repair lists should be discussed with the Chief Officer and the Second Engineering Officer and a drydock repair list made for the vessel.

Inspect the vessel's cathodic protection system and note what attention and maintenance it requires.

Check the maintenance and safety requirements of the cargo access equipment.

Carry out a similar check for the ship's cargo-handling equipment.

Inspect the pilot operation equipment and check that the maintenance standards and operational procedures are in accordance with the appropriate regulations.

Further reading

ARMSTRONG, M. C. *Pilot Ladder Safety* (International Maritime Press: Woollahra, Australia, 1979).

BUXTON, I. L. *Cargo Access Equipment for Merchant Ships* (Spon: London, 1978).

THOMAS, B. E. M. *Management of Shipboard Maintenance* (Stanford Maritime: London, 1980).

5

Oil Tankers: Cargo Operations

Pipeline systems

Pipelines are simply what their name suggests, lengths of steel pipes which connect groups of cargo tanks to one another and by which those tanks are loaded and discharged. Short lengths are bolted together by means of flanges or expansion joints. The latter consists of an oil-tight metal collar which surrounds the ends of two lengths and as the ends of the lengths do not touch, any horizontal thermal expansion or contraction will not damage the pipeline. Figure 5.1 shows a cross-section of an expansion joint. Lines pass directly through bulkheads, once again being secured by oil-tight flanges, and any sharp turns are constructed by bolting short curved lengths of pipe, known as bends, into the system. Branch lines are short lengths of pipes which serve individual tanks, the ends of such pipes expanding into a shape known as 'bellmouth', 'elephant's foot', or simply 'tank suction'. The tank main pipelines connect with cargo pumps and the deck pipelines, the diameter of the pipes varying from 25–30 cm (10–12 in) to 91 cm (36 in) depending upon the size of the vessel.

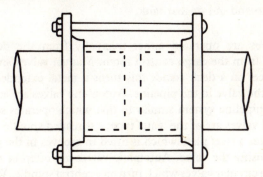

Figure 5.1   Expansion joint. For purposes of clarity only two bolts are shown

155

The various valves fitted within the pipeline system fall into five basic categories:

Manifold valves Cargo is loaded or discharged via shore hoses or metal loading arms which connect to athwartships deck pipelines known as manifolds. Valves which are constructed in the lines close to the connecting flanges route the cargo as desired by the ship's officers.

Drop valves Each of the main tank pipelines has a counterpart on deck and loading lines which lead vertically from the deck lines to the tank lines are known as drop lines. Each main line has one or two drop lines and drop valves control the flow of oil in those lines. Refer to Figure 5.2 and locate the drop valves and the other valves mentioned in this section.

Master valves At each place where a fore-and-aft pipeline passes through a tank bulkhead a valve is fitted in the line. This is known as a master valve and separates tanks served by the same fore-and-aft line. Tanks are usually constructed in sets of three transversely and are numbered from forward, e.g. the foremost three tanks are called 'One Port' (1P), 'One Centre' (1C), and 'One Starboard' (1S). The three tanks as a set are known as 'One Across' (1X). Thus to separate the cargo in 1X from 2X the master valves in the lines at the athwartships bulkhead between the tanks must be closed.

Crossover valves Athwartships tank lines joining the main lines are known as crossover lines and the crossover valves separate the main lines from each other as well as separating individual tanks. Thus 2P can be separated from 2C by a crossover valve. Briefly, master valves separate in a fore-and-aft direction and crossovers in athwartships direction.

Tank valves Close to each bellmouth is located a valve which controls the flow of oil into and out of that tank.

These valves are operated either manually from the deck above or automatically from the cargo control room. Manual valves are operated by turning a wheel on a deck stand; this turns a metal extended spindle rod which opens the valve in the pipeline. Automatic valves are activated by an hydraulic oil pipeline system similar to that which operates steering gears. Many manual valves are of the 'gate' type; a threaded spindle when turned, vertically moves a steel plate which is fitted in grooves in the pipeline, thus opening or closing the valve. Automatic valves are often of the 'butterfly' type; these are circular valves which turn on a central spindle. When open the plate is parallel to the pipe direction and the oil flows past it; when closed the

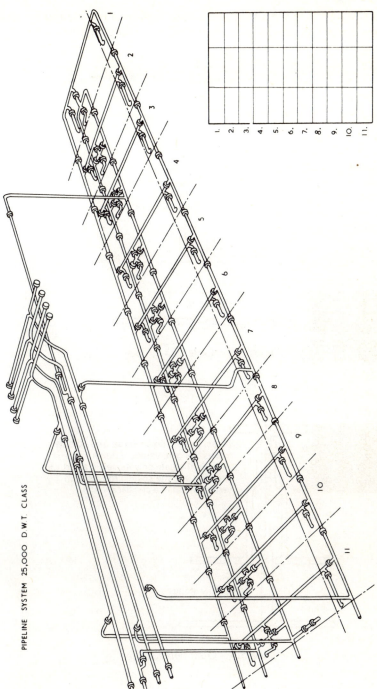

PIPELINE SYSTEM 25,000 D.W.T. CLASS

Figure 5.2 Cargo tank pipeline system of a product tanker

157

plate turns across the pipe, thus preventing the flow of oil. Figures 5.3 and 5.4 show the operation of a gate valve during a discharging operation.

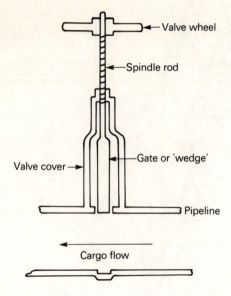

Figure 5.3   Gate valve, open

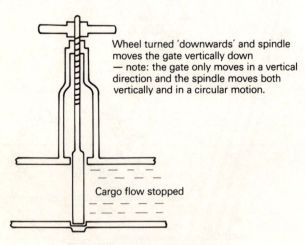

Figure 5.4   Gate valve, closed

158

Ring main system

This system is now found only on older ships. It has basically a 'square' or 'circular' layout whereby, if necessary, oil can be pumped up one side of the ship, across to the other, and then back down that side. On vessels with midship's pumprooms one ring main serves the tanks forward of the pumproom while another ring main serves the after tanks. On ships with an aft pumproom several layouts are common. One is shown in Figure 5.5

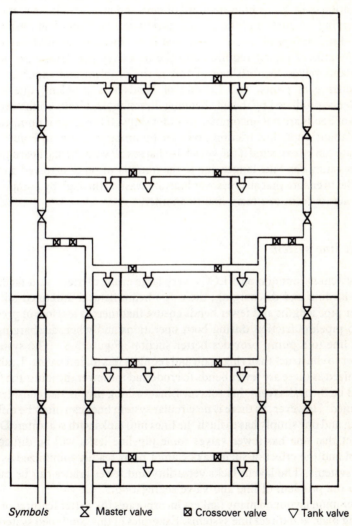

Symbols   ⋈ Master valve   ⊠ Crossover valve   ▽ Tank valve

Figure 5.5   Ring main system

whereby one main serves the forward tanks and the outer pumps and another the after tanks and inner pumps, both ring mains being connected with each other in three wings by short lengths of pipes.

Ring main systems are generally found on older product carriers ('products' refers to oils such as petrol which have been refined from crude oil) where versatility is required for the carrying of various grades of products. Versatility is the keyword for this system; with both ring mains being connected to each other any pump can discharge any tank. It also assists the loading and transporting of grades as one can ensure good pipeline segregation of cargo. A high standard of tank and line cleanliness is essential for carrying products and the circular layout aids tank and line washing. A major disadvantage of the system is that it is expensive to build due to the extra lengths of piping required and the necessary proliferation of joints, bends and valves. An offshoot of this is the problem and expense of the maintenance of joints, etc. Erosion of bends is a problem, due to the turbulence produced by the oil changing direction, and leaks on the external radius of bends are not uncommon on older ships. If the cargo is pumped by a 'roundabout' route line friction slows the pumping rate, e.g. in Figure 5.5 a slow rate can be expected if the vessel discharges 5P with the starboard pump. Line washing can take longer due to the number of pipes involved and it is essential to ensure that all crossover lines are washed through by pumping sea water across from one main line to another.

Direct line system

The system is common on VLCCs (very large crude carriers) as it facilitates quick loading and discharging, the cargo being natural unrefined oil. The shorter pipe lengths and fewer bends ensure that there is less loss of pressure due to pipeline friction during both operations and when discharging the direct line to a pump provides better suction (Figure 5.6). The system is cheaper to construct than ring main and requires less maintenance. Leaks are minimized as there are fewer bends to erode and the fewer the joints the fewer should be the leaks from that source. Line washing time is also considerably shortened. However, as there is no circular system lines can often be difficult to wash and one simply has to flush the lines into tanks with sea water. Due to the fact that one has fewer valves some pipeline leaks will be difficult to control and the effect of such leaks cannot be as readily minimized as with other systems. The layout lacks versatility and fewer grades can be carried due to the problem of line and valve segregation.

A common layout on many tankers incorporates the better features of both the ring main and direct line systems. Examples of this combined system can be seen in Figures 5.2 and 5.7.

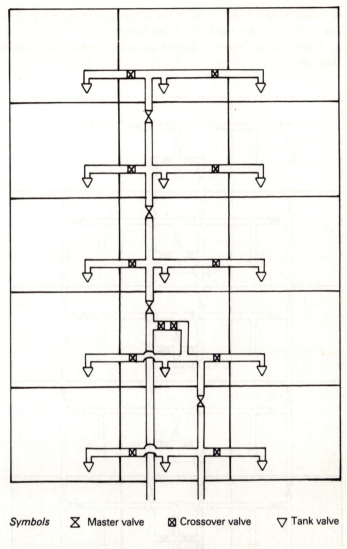

Symbols   ☒ Master valve   ⊠ Crossover valve   ▽ Tank valve

Figure 5.6   Direct line system

Free flow system

On some VLCCs the main pipeline is not used for discharging. Gate valves (sometimes known as sluice valves) are constructed in the tank bulkheads, and when these are opened the stern trim causes the oil to flow to the aftermost tanks where direct lines to the cargo pumps are located (Figure 5.8). This is a very fast method of discharging and the tanks are also efficiently drained as the large bulkhead sluice valves permit the oil residue to readily flow aft.

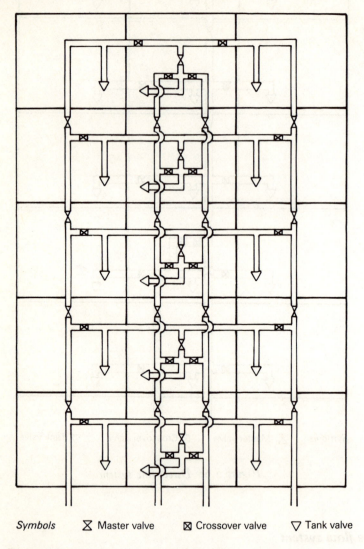

Symbols   ✕ Master valve   ⊠ Crossover valve   ▽ Tank valve

Figure 5.7   Combined system. Also known as the cruciform system

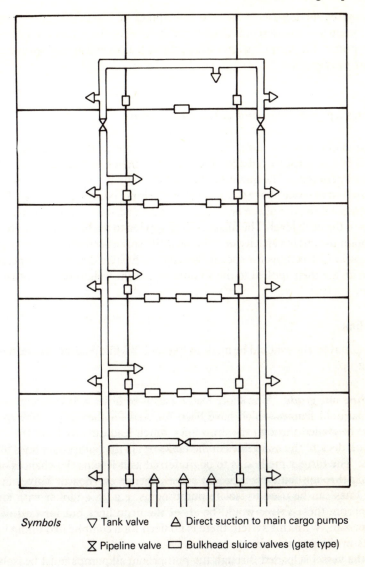

Symbols   ▽ Tank valve   △ Direct suction to main cargo pumps

✕ Pipeline valve ⬜ Bulkhead sluice valves (gate type)

Figure 5.8   Free flow system

Stripping lines

All the above layouts have stripping lines, separate from the cargo lines, incorporated into the systems. These are small pipelines, connected to low-capacity pumps, which are used for draining or stripping out the last few

163

centimetres of oil in the tanks. The oil stripped out is pumped to an aft cargo tank, known as the slop tank, and from there it is pumped ashore by a main cargo pump. The stripping lines and pumps are also used in tank washing and ballast operations.

Lining up pipelines and cargo operations

Cargo calculations are not dealt with in this chapter as there are several good tanker books already available which give instruction on that complex area of tanker operations. The quantity of cargo in a tank is obtained by measuring the amount of space above the oil level which is not occupied by cargo. This is known as taking the ullage or ullaging and is the distance from the top of the cargo to the deck level. The ullage is then applied to the tank calibration table to obtain the cubic capacity or volume of the space occupied by the cargo in that tank. Corrections for specific gravity (or relative density), temperature and trim are then applied to the volume to find the weight or quantity of the cargo.

Loading

Constant reference should be made to Figure 5.2 while reading this section on lining up.

Loading one grade The shore loading arms or hoses will be connected to the manifold flanges which have been selected for the cargo. The oil can either be loaded through the drop lines which lead directly from the cargo lines on deck to the main lines in the tanks or via the pumproom lines to the tanks. The former method is to be preferred as it lessens the chances of oil leaking through faulty pumproom valves, thence overboard. Valves in the deck lines can be used to isolate pumprooms, e.g. in a tanker with an aft pumproom these valves would be aft of the drop lines but forward of the pumproom. With this system there is no need for a complicated lining up of valves in the pumproom.

If the vessel is loaded through the pumproom all pumps must be isolated by closing the valves on the suction lines to them and also the valves on the deck delivery lines coming from the pumps (refer to Figure 5.9).

Hand-operated valves which are to be kept shut during loading should have their wheels lashed in the closed position. This serves the dual purpose of warning crew members not to open the valves and of preventing such valves 'walking back' or opening due to vibration or cargo pressure. Overboard sea discharge valves which are located in the pumproom are the valves through which the clean ballast is discharged prior to loading. When

164

de-ballasting is completed such valves *must* be lashed in the closed position. Usually the Shore Liaison Officer will not permit loading to commence until he has seen the lashings. The final line setting will now be carried out with all lines being made common so that the cargo can flow to any tank, the flow into each tank being controlled by that particular tank valve.

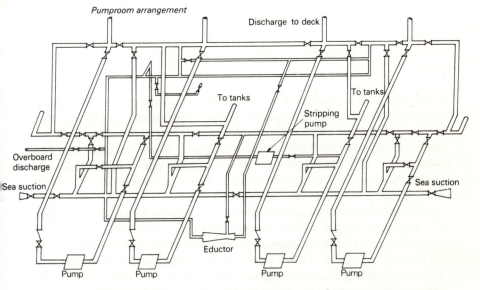

Figure 5.9   Pumproom pipeline system of a product tanker

When all the preparations have been completed and the appropriate tank valves opened, the manifold valves will be opened and loading commenced. It is general practice to commence loading into one or two tanks only, so as to ensure that the vessel has been lined up properly. A man should be stationed at the manifolds, ready to instruct the shore to stop loading in the event of a leaking connection or any other safety or cargo hazard. It is also common practice to load 'groups' of tanks, e.g. the wing tanks can be loaded first and then 'topped-up' during the loading of the centre tanks. The term 'topping-up' is used to describe the operation whereby the oil level in a tank is slowly brought up to its final ullage. Tanks should also be 'stepped down' to further spread out the time of the topping-up period, i.e. if it was possible to view a group of tanks from the side they should have the appearance of stairs (Figure 5.10). These procedures space out the topping-up operations and prevent 'panic stations' when quite a few tanks require topping-up at the same time. Accidents sometimes occur when changing over from full tanks to empty ones. Empty tanks must always be partially opened some time prior to the changeover, otherwise cargo-loading pressure may prevent valves opening or closing and thus causing a spill.

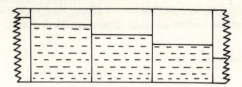

Figure 5.10 Stepping down oil cargo tanks

Loading more than one grade The number of grades carried will depend very much on the design of the vessel. Many ships can carry only two or three grades but product carriers are usually multi-grade. Product carriers are often divided into two classes, those which carry 'white' or 'clean' oils and those which carry 'black', 'dirty' or grades of crude oils. White or clean oils are highly refined products such as motor and aviation spirits and black or dirty products are diesel, fuel and furnace oil.

The ideal arrangement is to have a separate pipeline and a separate pump for each grade; this is seldom possible. However, compatibility charts are available which give instructions as to whether certain grades may be loaded through the same line. Thus some slight contamination is permissible and several grades may be loaded via one line. Any such instructions must be adhered to as consignees are very strict with regard to contamination standards. There should be a two-valve separation, if possible, between different grades.

Bulkhead leakage is a problem when carrying grades. When carrying incompatible grades the one most likely to be adversely affected by mixing should be loaded to a higher level than the less vulnerable grades in adjoining compartments. Thus diesel oil should be loaded to a higher level than furnace oil (both black oils); any leak of diesel would enhance the furnace oil. Similarly, kerosene (paraffin) should be loaded higher than motor spirit; although both are white oils, any contamination of kerosene by motor spirit could be disastrous for an unfortunate householder or camper.

Discharging

While the shore hoses are being connected the tanks should be ullaged, water dips, temperatures and samples taken, and the pumps warmed through. Water dips ascertain the amount of water, if any, which has settled out of the cargo during transit. A tank is 'sounded' by manually lowering a metal rod with a paste smeared upon it to the bottom of the tank. If water is present the colour of the paste changes, thus indicating the water level. Tank calibration tables convert this figure to the amount of water in a tank.

The pipeline suction valves to the pumps and on the deck delivery lines should be opened, leaving the tank and manifold valves to be opened just before discharging commences. Once again it is a safe practice to commence discharging from one tank only in order to ensure that the ship and shore lines have been set properly.

Sometimes more than one grade will have to be discharged using the same pump. Some contamination will result but by working with the terminal representative this can be kept to an acceptable minimum. As above, the vulnerable grade should be discharged first. If a pump is to be used for diesel oil and marine fuel, then the diesel should be discharged first.

A discharge plan must be adopted which will allow the main pumps to work for as long as possible while permitting the stripping pumps to drain some tanks when there is still a considerable amount of cargo in other tanks. Most of the tanks should already have been drained before the final tanks are pumped dry. A good stern trim must be maintained during the discharge in order to assist tank stripping.

Ballasting

It is best to ballast the vessel when the discharge is completed. However, it may be necessary in exposed locations to ballast during discharge. The tanks to be ballasted must be completely isolated from the cargo and a two-valve separation should be regarded as a minimum separation in any pipeline which contains both water and oil. It is sometimes possible to commence ballast without using the pumps, i.e. 'running in' the sea water. This can result in oil escaping overboard and the author would not use that method. Cargo pumps used for ballasting should be started at a slow rate before opening the sea valves.

At some terminals de-ballasting is concurrent with loading; extreme care must be taken during the operation. During the voyage clean ballast will have been pumped into some cargo tanks and all pumps and lines will have been thoroughly cleaned. Part of the ballast will be run out before berthing, leaving only sufficient on board to keep the vessel manageable. At the end of the de-ballasting operation the tanks will have been fully drained and the main cargo pumps and lines will also have been stripped.

Under current regulations some ships will have permanent clean ballast tanks. These are known as segregated tanks and are served by pumps and a line system which are entirely separate from cargo pumps and lines. Cargo and segregated ballast systems are usually worked simultaneously.

General precautions during cargo operations

Adherence to safety procedures is of paramount importance on tankers; your life depends on such adherence. Much of the rest of this chapter will be concerned with safety. Make safety a way of life or your career could be a way of death.

The Chief Officer should have a detailed cargo operation schedule or plan available so that personnel are fully aware of what exactly is happening. Officers must be conversant with the maximum loading rates of cargo. For some cargoes the rate is limited for intrinsic safety reasons. However, the ship generally stipulates the rate; this should not exceed the capacity of the ship's vapour lines to clear cargo gases or inert gas. Loading or discharging rates should be agreed between ship and shore. Pipelines and hoses should never be over pressurized. Cargo-handling signals must be agreed and understood between ship and shore personnel. Arrangements for indicating the following must be made:

(a) Stand By
(b) Commence Operation
(c) Slow Down
(d) Stop Loading or Discharging
(e) Emergency Stop

The loading rate must never be controlled by operating the manifold valves but should be reduced by shore personnel.

Manual ullaging and tank sampling should be carried out only through the sighting ports constructed in the tank lids and these should only be opened long enough to do the job. Personnel should stand at right angles to the wind direction to avoid inhaling gas. Manual steel ullage tapes and metal sampling cans should never be used while a distillate (clean) oil is being loaded. This is to avoid the risk of sparks since the distillate oil may be statically charged and non-conductive ullage tapes, etc. must be used. Gauges which are fixed to the ship's structure, such as 'Whessoe', are considered to be non-conductive. When loading, sufficient ullage should be left to allow for the expansion of cargo which is being transported to warmer climes.

It should be remembered that bunkers are often taken during cargo operations. The appropriate safety and stability precautions should be observed.

Procedures for line and hose clearance will depend upon the shore facilities available, but whatever method is used all precautions against spillage must be taken.

Attention should be given to the tank pressure relief valves but these will be considered later.

Comply strictly with the terminal 'Fire Notice' which gives details of emergency procedures in the event of fire and with the 'Ship/Shore Safety Check List'.

Ship/Shore Safety Check List

This is a joint list provided by the terminal and which is signed by a ship and a terminal representative. It states the general precautions required before loading, discharging, ballasting, tank cleaning and gas freeing when alongside and should be adapted for use when any of the above operations are carried out at sea. An example of such a list is given below.

1 Appropriate personnel notified that cargo operations are about to commence.
2 Sufficient personnel available to maintain an efficient deck watch and an anti-pollution patrol.
3 Warning notices displayed, e.g.
   Warning
   No Naked Lights
   No Smoking
   No Unauthorized Persons
4 Fire appliances ready for immediate use.
5 Moorings tight and emergency towing wires correctly positioned.
6 Agreed ship/shore communication systems working.
7 No unauthorized persons on board.
8 No unauthorized work to be carried out.
9 No unauthorized craft alongside.
10 No naked lights and no smoking unless in designated areas.
11 Safe lighting available, e.g. gas-tight torches, etc.
12 Galley precautions observed.
13 Cargo tank lids closed.
14 Sea valves and overboard discharge valves when not in use closed and lashed.
15 Cargo lines properly set. All valves not in use closed and lashed.
16 Shore connections properly secured and supported.
17 Manifold drip trays in use.
18 Tank venting system checked and properly set.
19 Tanks inerted and inert gas system checked.
20 All doors and ports to accommodation closed and any ventilators suitably trimmed.
21 Air conditioning unit on internal air re-cycling.
22 All necessary spark arrestors in good condition and in place.
23 Safe ship/shore access.

24 Ship ready to move under own power.
25 Emergency shut-down procedure understood.
26 Portable radio transceivers which are used for communication to be of an approved type.
27 Ship's main radio aerials earthed.
28 Appropriate flag and light signals to be shown.
29 All deck scuppers plugged to prevent oil leaks overboard.

Safety guides

The International Chamber of Shipping (ICS) publishes a booklet, *Safety in Oil Tankers*, which contains a synopsis of safety hazards and procedures. It should be read.

The tanker officer's companion, though, should be the *International Safety Guide for Oil Tankers and Terminals* (ISGOTT). Published by the ICS and the Oil Companies International Marine Forum (OCIMF), this guide replaces the ICS *Tanker Safety Guide (Petroleum)* and the OCIMF *International Oil Tanker and Terminal Safety Guide*. The contents of the guide are in two parts:

Part 1—Operations   This covers operational procedures and is designed to provide guidance in safe practices. This guide has a more practical approach than the earlier guides; it has been produced by and for operators who are aware of short cuts that are taken, the hazards that are encountered, and the fact that most accidents are caused by human error.

Part 2—Technical Information   Contains additional information, mainly the theory behind the procedures in Part 1.

The general scope of the guide is to make recommendations for practices to be adopted by tanker and terminal personnel to ensure safety in operations relating to the carriage by sea and the handling at terminals of crude oil and petroleum products.

The basic approach of Part 1 has been to arrange the material so that each chapter is concerned with a particular type of operation. However, some chapters deal with precautions that are generally applicable and these should be followed as well as those for the operation concerned. Each chapter has a small introductory paragraph describing the scope of its contents and, where appropriate, drawing attention to related chapters.

The guide is an excellent publication and is much more practical and seamanlike than many of the plethora of publications which have inundated seamen in recent years. Officers serving on tankers *must* have a working knowledge of its contents. The first two chapters deal with the general hazards of petroleum and precautions on tankers. Port procedures, precautions at a berth, shore liaison, cargo and ballast handling, tank cleaning and gas freeing, etc., are dealt with in subsequent chapters. Due to recent accidents particular attention should be paid to:

Chapter 9: Fixed Inert Gas Systems
Chapter 10: Entry Into Enclosed Spaces

Useful information is also given on combination carriers, packaged goods and, in particular, Chapter 13: Emergency Procedures. The chapter number would not appear to mollify the anxieties of superstitious seamen.

The guide has received international recognition from IMO, and has been accepted worldwide as the standard safety manual for the tanker and oil industries. However, the guide received some slight criticism in a report from the Tanker Accident Working Group (November 1981). The Group found 'that some of the recommended safety precautions have either been ignored or misunderstood' and that 'in some respects ISGOTT is not sufficiently explicit'. I have no doubt that these statements are true but safety lies mainly in the hands of the operators. If precautions are ignored the end results are *accidents!*

Main sources of ignition on tankers

It should be pointed out that oil does not burn. The main problem is the gas which evaporates from the oil mixing with air to produce a flammable mixture. Thus we must avoid igniting the mixture by paying attention to the sources enumerated below.

Smoking

Secret smoking is dangerous and a policy of controlled smoking is advisable. Smoking should be banned in areas such as pumprooms, stores and workshops but approved in specially designated locations such as mess rooms. Care must be taken in selecting these areas with regard to ventilation intakes, etc.

Electrical equipment

Every piece of electrical equipment is a fire hazard, especially generators and motors where arcing can occur at brushes, etc. This equipment is usually found in accommodation and work spaces, so care must be taken to prevent the entry of gas into such spaces. Equipment such as torches, and tank-cleaning and gas-freeing appliances must be of approved design and must not be defective.

Metal

Metal striking metal can cause an incendive spark, so one should be wary of using chipping hammers, etc. Never carry out work when flammable gas may be present. Non-ferrous tools, also known as 'non-sparking' tools, can be hazardous; when these tools impact with iron oxide, a very rapid reaction can take place between the metal tool and the oxygen in the iron oxide resulting in a thermite spark. I would not advise the use of these tools as it gives people a false sense of safety.

Domestic equipment

Electric shavers, cooking appliances, music centres, etc. can ignite flammable vapours. Keep an eye open for the transistor mister (or sister) sunbathing on deck while listening to a radio!

Aluminium

Aluminium equipment knocking against rust can cause a heat (thermite) flash. Similarly, aluminium paint over rust may generate heat if struck by an object. Never use aluminium paint in pumprooms, etc, and beware of all equipment (including shovels) made of alloy.

The galley

Check whether the galley appliances present fire hazards and note any port regulation regarding galleys.

Lightning

The most likely place to be struck is at the top of a vent riser on a mast, thereby igniting tank vapour. Due to the use of inert gas this hazard has been considerably reduced. One should note, though, that a recent incident has occurred when a ship with a non-operational system was struck by lightning, causing death and the ultimate scrapping of the vessel.

Spontaneous combustion

Oily or paint-soaked rags and waste will generate heat if left in a pile. Keep storerooms clean.

Sparks

Funnel sparks, especially when manoeuvring, can be hazardous. Make sure that all anti-spark gauzes or flame screens are kept in excellent condition.

172

Bunker barges can also produce sparks especially if one observes, as I have, a bunker barge crew member lighting a cigarette on deck!

Tank anodes

Zinc anodes should be the only type to be fitted on tankers as magnesium anodes can produce incendive sparks and aluminium anodes can produce thermite sparks if both are dropped from a height. Zinc anodes are free from such a hazard.

Ship to shore electric currents

The ship and the shore form two electrodes of a very large electrolytic cell due mainly to the differences in the electrolytic environment of the two structures. If the shore cargo arm is all metallic, it provides a very low resistance connection between ship and shore and an incendive arc can occur if the large current is broken during disconnecting operations. This is avoided by inserting an insulating flange within the length of the arm or at the flexible hose connection to the shore pipeline.

Static electricity

This is a major hazard and requires a section devoted solely to it.

Static electricity

Definition: the electric charge produced on dissimilar materials through physical contact and separation. Certain operations produce accumulations of electric charge which may be released suddenly in electrostatic discharges, thus producing a spark which can ignite a flammable vapour. When two dissimilar materials come into contact, charge separation can occur at the interface, i.e. one material becomes positive and the other negative. If the materials are separated the energy expenditure in pulling the charges apart appears as a voltage between the separate charges. If a convenient path presents itself, the rejoining of the charges results in an electric current which, if the voltage is substantial, may be sufficient to break down the atmosphere and cause a rapid discharge in the form of a spark. Similarly, when charges are separated, a voltage distribution can be set up throughout the neighbouring space, e.g. a cargo tank, and this is known as an electrostatic field.

Some causes of static charging are given below. The word 'petroleum' should be taken to mean crude oil and liquid products derived from it. When petroleum flows through a pipeline, a preponderance of molecules of one sign occurs at the interface between the petroleum and the pipe, and the oil becomes charged.

Loading overall

There is some confusion concerning this term. I have heard it being used to describe the operation of loading all tanks at the same time. The ISGOTT definition states that it is 'the loading of cargo or ballast "over the top" through an open ended pipe or by means of an open ended hose entering a tank through a hatch or other deck opening, resulting in the free fall of liquid'. Lubricating oils are sometimes loaded by the latter method but whether one loads over the top or all tanks at the same time, both methods deliver charged petroleum (caused by pipeline flow) throughout the vessel. Loading in this manner also entraps air in the oil in the tank, which bubbles back to the surface causing a charged spray. In addition to static hazards the general turbulence involved increases cargo evaporation, thus causing high vapour concentrations in the ullage space.

Settlement of water through petroleum

An electric double layer occurs at the water/petroleum interface. The relative motion of the water, and petroleum of low conductivity, will cause separation of the outer and inner charged layers and a voltage will build up. This can occur during tank washing and is a hazard particularly associated with slop tanks.

Splashing

When petroleum is splashed on another material each droplet leaves a film of liquid adhering to the surface of the material. This film will contain the outer layer of the electric double layer while the droplet itself will carry away the inner layer causing the droplet to become charged. If the droplets, in the form of spray, settle on an unbonded conductor a large charge can accumulate and can result in a spark discharge to a bonded structure.

Air release in cargo tanks

The passage of air through oil does not generate static electricity but when the bubbles of air burst at the surface they give rise to a spray of liquid particles which may become charged as in splashing.

Steam

Water droplets issuing at high velocity through a nozzle become charged (similar to petroleum flow in a pipe) and this can produce a charged mist.

This is a hazard associated with tank washing using hot water or when steaming a tank.

Inert gas

A similar process to that described for steam can apply to high-velocity discharge of inert gas into a tank. This applies in particular to carbon dioxide when charged frozen particles are formed due to the rapid cooling which takes place.

Precautions to prevent ignition due to static charging

1 The following must be made of non-conductive material:
   (a) hand-held ullage tapes
   (b) ullage sticks
   (c) sampling containers
   Polypropylene ropes can accumulate charges and should never be used to lower sampling containers, etc. into tanks. Use natural fibre lines.
2 During tank washing never introduce metal objects into the tank other than grounded washing machines and do not disconnect tank hoses from their hydrants until the hoses have been pulled out of the tank.
3 Water in suspension in petroleum produces a high static charge which is most likely to be present at the beginning of loading. A slow initial rate will reduce the static.
4 Do not load overall.
5 Do not load white oil into many tanks at the same time.
6 After discharge do not ballast a white oil tank until it has been thoroughly stripped.
7 No unearthed conductor capable of collecting a charge should be inserted into a tank being loaded.
8 It is recommended that at least thirty minutes must elapse after completion of loading before hand-held or other conductive equipment is used. The author would recommend that such equipment not be used under any circumstances.
9 Compressed air or inert gas should not be used for line clearance unless precautions are taken to prevent the air or gas entering the tank.

Cargo pumps

There are two basic types of cargo pumps on oil tankers, positive displacement and centrifugal, both of which are usually driven by steam.

Reciprocating positive displacement pumps

This type of pump moves a certain amount of liquid with each pump cycle. The pump piston draws liquid through a non-return suction valve into a cylinder which is known as a 'bucket'. The cylinder is full at the end of the suction stroke and on the reverse stroke the liquid is expelled from the cylinder through a non-return discharge valve. Most pumps are 'double-acting' to ensure a steady flow of oil, i.e. the piston and cylinder are arranged so that whether the piston is moving up or down a flow of oil comes from the pump. Most pumps of this type are duplex, having two buckets and two pistons, which ensures that suction is not lost at the end of a stroke. The drive unit is an integral part of the pump, such pumps being situated in the pump room.

Positive displacement pumps move a low volume of oil at relatively high pressure. Their use on tankers is generally restricted to stripping pumps. A typical stripping pump on a VLCC would move 400 tonnes of oil each hour at a working pressure of approximately 100 psi or 7 kg/cm^2.

Centrifugal pumps

An impeller, which is inside a casing, physically moves the oil by means of a 'throwing' movement which is similar to the expelling of water from a bicycle tyre when cycling in wet weather. The oil is 'sucked' into the casing via a suction valve from the tank main lines and is pumped to the deck lines via a discharge valve. The pump provides a continuous flow of oil and it is powered by a steam turbine drive unit which, for safety reasons, is installed in the engine room. The impeller is turned by means of an extended rotating shaft which penetrates the engine room/pump room bulkhead through a gas-tight seal.

Centrifugal pumps move large volumes of liquid at a relatively low pressure and consequently are generally used as main cargo pumps (MCPs). A typical main cargo pump on a VLCC can move 4,000 tonnes of oil each hour with the pump running at 1,400 revolutions per minute at an operating pressure of approximately 150 psi or 10.5 kg/cm^2.

Pumproom routines

A mud box is usually located forward of the pump suction valve and it should be opened and inspected at frequent intervals. Particular attention should be paid to the perforations in any strainer plate. Loose nuts and similar objects are occasionally found in strainers and mud boxes. Valve glands, pipe joints and flanges should be checked for tightness and sharp bends in piping should be inspected when possible for signs of internal erosion. The nuts on any inspection plates should be checked for tightness and all pumps should be given a visual examination before being operated. Strict safety precautions should be adhered to before carrying out any pumproom maintenance.

The build-up of an explosive air/hydrocarbon gas mixture during cargo pumping operations must not be permitted and precautions must also be taken to prevent a heat source developing. Inspect all pipe joints, valve glands, pump glands, and pump seals for oil leaks. Ensure that all pump bearings are correctly lubricated and check pump casings to ensure that the pumps are not overheating. Check the mechanical seals through which a pump shaft penetrates a pump casing to ensure that the seals are not leaking and that they remain cool. If a pump produces excessive noise or vibration it should be stopped and an engineer should inspect it.

All main cargo pumps have an 'Emergency Shut Down' device, often known as the 'Trip', and all officers should be conversant with emergency shut down procedures.

Cavitation

Cavitation occurs in a centrifugal pump when the impeller rotates without actually transferring a liquid. The impeller may also rotate in a partial vacuum due to a low oil flow or in an oil/gas mixture. The gas may be vapour from the cargo or inert gas from the tank space above the cargo level, hence the term 'gassing-up'.

Gassing-up can occur when the oil level in the tanks is low. The actual discharge rate of a pump is affected by its net positive suction head (NPSH), i.e. the height of the cargo above the centre of the pump below which the pump will not operate properly. Thus when cargo is low in a tank the pump cannot discharge at a high flow rate and may cavitate. If the pump is 'running' too fast a whirlpool effect may be produced around the tank suction which permits vapours to be drawn into the pump and gassing-up will occur. Many tanker officers use the terms 'cavitation' and 'gassing-up' as interchangeable terms for one phenomenon.

The term 'volatility' refers to the tendency for a liquid to produce gas by evaporation. Cargoes with high volatility such as naphtha are often difficult to discharge as the pumps have a tendency to gas-up throughout the operation. The term 'viscosity' refers to the resistance of a fluid to shear forces and hence to flow. It is possible to obtain a good flow or discharge rate with cargoes of medium viscosity as the impellers can get a firm 'grip' on the oil. There is a tendency for pumps to cavitate when discharging cargoes of low viscosity as impellers may rotate without transferring the oil.

If cavitation occurs and suction is lost, a centrifugal pump will overspeed

177

and may be damaged if the situation is not rectified immediately. When the oil level in a tank falls the speed of the pump discharging that tank should be slowed accordingly. The reduced flow rate should prevent excessive turbulence within the pump and minimize the whirlpool effect in the tank. It may be necessary to partially close the tank suction valve to further reduce the flow rate from that tank. The discharge valve from a centrifugal pump may also be partially closed as the action of the pump pushing against the increased discharge head or pressure will reduce cavitation. The manifold valves should never be partially closed in an attempt to reduce cavitation.

If a pump is discharging a set of tanks, one full tank should be kept in reserve and when the cargo in the other tanks falls to a low level the reserve tank should be opened to ensure a good flow rate. Thus by judicious use of the back-up tank the other tanks can be discharged by a main cargo pump to a very low level. This reduces the amount of cargo which the stripping pump must transfer and therefore reduces the overall stripping time. Most tankers have three or four main cargo pumps and it is important that all the centrifugal pumps are operated at the same speed. If one pump runs at a slower speed than the others it will not transfer the cargo and will thus cavitate and overheat. Pump seals and glands should be inspected for air leaks into the pump as such leaks will increase the possibility of cavitation.

Some vessels have a 'vac-strip' system incorporated with the cargo pumping system in the pump room. Air and vapour are prevented from entering the cargo pump by the installation of a small separator tank on the suction side of the pump. In the tank air and vapour are automatically separated from the oil before the cargo reaches the pump and thus the pumping performance is not affected. The 'vac-strip' system also automatically controls the speed of the pump so that it remains compatible with the amount of cargo reaching the pump.

Cargo operations when not secured alongside

Tankers frequently load or discharge at single buoy moorings (SBMs). The vessel is secured from the focsle head to a large buoy and a floating hose for transferring the cargo is hoisted by the ship's lifting appliances and connected to the manifold. Tankers may also work cargo at berths in which the securing arrangements consist of the vessel's anchors forward and a number of wires to buoys aft. Lightering operations are sometimes carried out whereby the discharging vessel is anchored and a specialized service ship is secured alongside to receive a cargo. The procedures for conducting the latter operation, known an a 'ship to ship (STS) transfer operation', are found in the ICS/OCIMF guide *Ship to Ship Transfer Guide (Petroleum)*.

A high operational and safety standard is maintained during STS transfer operations and spills seldom occur. However, if a spill does occur in British waters the following authorities are involved.

1 The Department of Transport has the overall responsibility for dealing with the incident and the department has a 'Marine Pollution Control Unit' (MPCU) on twenty-four hour availability at various locations around the United Kingdom.
2 HM Coastguard is in control of communications and is responsible for alerting the various authorities.
3 The major oil companies have oil spill units with dispersant spraying vessels. These vessels work in close liaison with the Department of Transport controlled dispersant spraying tugs and planes.
4 Oil which reaches the shore is the responsibility of the local authority.

Procedures if a spill occurs when discharging at anchor

The procedures will vary slightly from country to country and the emergency procedures should be discussed with the local liaison officer. Some areas prohibit the use of the ship's dispersant chemical while other areas require the ship to cover the adjacent sea area with a foam blanket to reduce the possibility of a spark igniting the cargo vapour. The following procedures should therefore be considered as guidelines only:

1 The discharging vessel should immediately operate the emergency cargo shut-down procedure.
2 Inform the Captain and sound the Emergency Squad alarm.
3 Sound or activate any pre-arranged accidental spill signal (e.g. to port authorities or other ships).
4 Locate the emission site and attempt to stop the oil emission.
5 Notify the appropriate authorities.
6 Spray the area with locally approved dispersant with the permission of the Control Unit.
7 If permission is given, spray the sea with foam to reduce the fire and explosion risk.
8 The main engines should be available but inform the engineers as to the nature of the emergency.
9 Instruct an officer to keep a note of all activities together with the times of such activities.
10 If oil has been spilled on deck organize some of the crew in a cleaning-up operation.
11 The following information should be collated:
   (a) the estimated amount of oil spilled
   (b) the type of oil
   (c) the nature of the oil

    (d) if an oil boom surrounds the ship or not
    (e) the weather, sea, and tidal conditions
    (f) the position of the ship and the distance from shore

When time permits entries should be made in the 'Official Log' and the 'Oil Record Book'. Permission should be obtained before resuming cargo operations.

Oil record books

M1089 'Requirements for ships to keep oil records' should be studied in full. The notice reminds mariners that oil tankers registered in the United Kingdom must carry an oil record book in which details of the following operations or incidents must be recorded:

1 The loading, discharge, and voyage transfers of oil cargo, cleaning and ballasting of cargo tanks, discharge of dirty ballast and water from the slop tank, discharge of oily bilge water from machinery spaces and the pump room whether in port or at sea, and the disposal of oil residues.
2 Any occasion when oil or an oily mixture has been discharged in order to maintain the safety of a ship or to prevent damage to any vessel or cargo or to save life.
3 Any occasion when oil or an oily mixture has been discovered escaping, or to have escaped, from a ship due to damage or to leakage from that ship.

In addition, the Master of every vessel which engages in an oil transfer operation within United Kingdom waters must keep a record of the following particulars:
(a) the name and port of registry of the vessel
(b) the date and time of the transfer
(c) the place of the transfer
(d) the amount and description of oil transferred
(e) from what vessel or place on land, and to what vessel or place the oil was transferred
The record of each operation must be signed and dated by the Master.

    Full details of oil record books and oil transfer operations are found in the Prevention of Oil Pollution Act 1971, Oil in Navigable Waters (Records) Regulations 1972, and Oil in Navigable Waters (Transfer Records) Regulations 1957.

    Under the provisions of the International Convention for the Prevention of Pollution from Ships 1973 (MARPOL 73), as modified by the Protocol of 1978, which came into force on 2 October 1983, every oil tanker of 150 tons gross tonnage and above and every ship of 400 tons gross tonnage and above must be provided with an oil record book. Regulation 20 of Annex 1 to MARPOL 73 states the occasions when records must be made and Appendix III of Annex I to MARPOL 73 states the required 'layout' or 'form' of the oil record book. The Protocol of 1978 enumerates the supplements which must be added to the oil record book.

The Merchant Shipping (Prevention of Oil Pollution) Regulations 1983

The following M notices should be consulted with reference to the above regulations.

M1076. This contains guidelines for the conduct of initial, mandatory annual, intermediate and renewal surveys (required by MARPOL 73/78). The UK conducts annual surveys in substitution for unscheduled inspections of ships requiring an International Oil Pollution Prevention Certificate.

M1077. The notice defines oil as 'petroleum in any form including crude oil, fuel oil, sludge, oil refuse and refined products other than petrochemicals listed in a merchant shipping notice'. The notice lists such petrochemicals.

It should be noted that the above regulations apply to ships other than tankers as well as to tankers.

6

Oil Tankers: Routine Operations

Inert gas system

Regulations

For United Kingdom ships the requirements for inert gas systems are found in the following regulations:

> The Merchant Shipping (Fire Protection) (Ships Built Before 25 May 1980) Regulations, which became operative on 12 August 1985 and which apply to ships built before 25 May 1980
> The Merchant Shipping (Fire Appliances) Regulations 1980, as subsequently amended, which apply to ships built on or after 25 May 1980
> The Merchant Shipping (Fire Protection) Regulations 1984, which apply to ships built on or after 1 September 1984

The 1978 Protocol to SOLAS came into force on 1 May 1981. By the requirements of that Protocol all existing tankers of 70,000 metric tons deadweight and upwards had to have an inert gas system by 1 May 1983; existing crude carriers of 20,000 metric tons deadweight and upwards, and existing product carriers of 40,000 metric tons deadweight and upwards had to have an inert gas system by 1 May 1985. 'Existing tanker' for the purpose of the Protocol refers to ships which were in existence on the date of entry into force of the Protocol. All new tankers of 20,000 metric tons deadweight and upwards shall have an inert gas system. 'New tanker' for the purposes of the Protocol means a vessel for which the contract was placed after 1 June 1979, or was in an early stage of building on 1 January 1980, or which was delivered after 1 June 1982. Inert gas systems are mandatory on all tankers using crude oil washing and by May 1985 all product carriers of 20,000 metric tons deadweight and upwards fitted with tank washing machines greater than 60 m³/hour were also required to have inert gas systems.

Manuals

Students should refer to *Inert Gas Systems for Ships Carrying Petroleum*

182

Products in Bulk published by IMO, and *Inert Flue Gas Safety Guide* published by ICS and OCIMF.

Principle of inert gas system (IGS)

Three factors contribute to an explosion within a cargo tank:

1 Hydrocarbon gas.
2 Oxygen in sufficient quantity to support combustion.
3 An ignition source.

To prevent an explosion it is necessary to obviate at least one of the factors. The introduction of inert gas into a cargo tank reduces the oxygen content to a low level and also reduces the hydrocarbon gas concentration in the atmosphere to a safe proportion. Thus two factors have been made innocuous and protection against a tank explosion has been achieved.

Benefits of inert gas

1 It prevents explosions in cargo compartments.
2 It can be used as an extinguishing agent and thus can be installed as a fixed fire-fighting system.
3 Inerted tanks are maintained at a slight positive pressure. During discharging operations this pressure slightly increases the discharge rate.
4 The presence of inert gas in cargo tanks reduces the loss of cargo due to evaporation.
5 Inert gas in a cargo tank makes it possible to transport certain cargoes which react when mixed with oxygen, e.g. certain chemicals which can be contaminated by oxygen or which have a violent reaction to oxygen.
6 The installation of an inert gas system complies with certain mandatory regulations which then makes it possible for a ship to use the crude oil washing technique.
7 The reduction of oxygen content to a certain proportion in a tank atmosphere also causes a reduction of the corrosion process.

Ordinary air contains about 21 percent oxygen. Combustion can occur in a cargo tank if an air/hydrocarbon mixture exists and the gas concentration is within the flammable range. If the oxygen content of the tank is reduced to 11 percent combustion cannot occur regardless of whatever the hydrocarbon concentration might be. To provide a good margin of safety inert gas systems are required to maintain tank atmospheres at an oxygen content not exceeding 8 percent by volume. Inert gas systems on British ships are also required to 'be capable of delivering inert gas with an oxygen content of not more than 5 percent by volume' to the supply main to the cargo tanks.

The figure of 5 percent oxygen content is intended primarily to be a safety

183

standard but there is a marked reduction in tank corrosion once that proportion is attained.

The purpose or function of an inert gas system is therefore to reduce the oxygen content of cargo compartments by the introduction of inert gas into those compartments.

The gas used is the ship's funnel gas which is cleaned, cooled and distributed to the cargo tanks.

Components of inert gas plant

Reference should be made to Figure 6.1 which shows a schematic diagram of an inert gas plant.

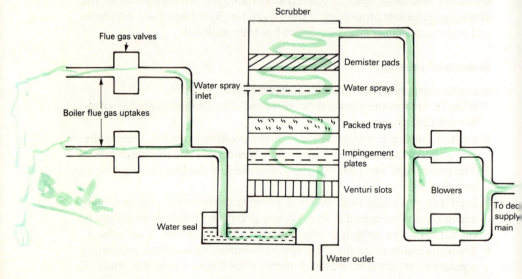

Figure 6.1   Inert gas plant

Inert gas source   The boilers in the ship's engine room are the source for the funnel gas used in the inert gas process. The gas is led from the boilers to the funnel via flue gas uptakes.

Flue gas isolating valves   These valves are located in the pipes which permit the hot, dirty flue gases to be delivered to the scrubber. The valves isolate the scrubber from the flue gas uptakes when the plant is not in use. If the boiler gas is not required for the inert gas process it flows up the funnel piping and is expelled to the atmosphere.

Scrubber   The scrubber has a three-fold purpose:

184

(a) to cool the flue gas

(b) to remove most of the sulphur dioxide

(c) to remove most of the soot particles

The designs of scrubbers vary considerably but in all scrubbers flue gas is brought into contact with large quantities of sea water by which the gas is cooled and cleaned. Before entering the scrubbing tower the gas receives an initial cooling by being passed through a water spray or by being bubbled through a water seal or 'trap'. The trap also prevents any flue gas which may leak through the isolating valves entering the scrubbing tower so that maintenance can be carried out in safety.

The tower can be thought of as a sort of sandwich layer cake made up of the following components (from bottom upwards):

(a) venturi nozzles and slots

(b) perforated impingement plates

(c) trays of packed stone or plastic chippings

(d) water sprays

The flue gas enters the bottom of the tower and moves upwards through down-flowing sea water, the layered arrangements ensuring maximum contact between water and gas.

Demister trays, which consist of polypropylene 'mattresses', are located at the top of the tower to remove water droplets from the gas.

The sea water which cleans the gas becomes acidic and is led overboard from the bottom of the scrubber by corrosion-resistant piping.

Blowers These are electric motor driven centrifugal blowers which deliver the clean, cool inert gas to the cargo tanks via the distribution system.

Components of inert gas distribution system

Reference should be made to Figure 6.2 which shows a schematic diagram of the distribution system. The gas is distributed through a pipe in which are fitted pressure control arrangements to regulate the flow of gas to the main distribution pipe on deck and to prevent any backflow of gas in the event of a mechanical failure in the plant.

Gas pressure regulating valve This valve is usually part of an automatic re-circulating unit whereby gas which is not required in the cargo tanks is re-routed back to the scrubber.

Deck water seal This is the principal barrier against backflow. In the basic type of seal, the 'wet seal', the inert gas bubbles through water from a submerged inlet pipe. If the pressure of the gas in the cargo tanks exceeds the pressure of the gas in the inlet pipe, water is forced up that pipe and any

185

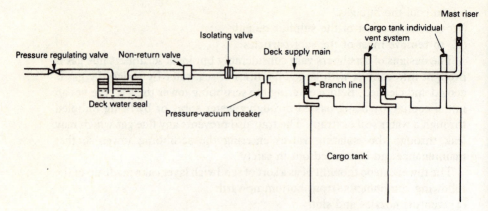

Figure 6.2   Inert gas distribution system

backflow is thus prevented. The disadvantage with the wet seal is that the gas must pass through a demister pad (which is within the seal unit) to reduce the possibility of water droplets being carried over.

Deck non-return valve   A mechanical valve which also prevents backflow but in addition prevents any cargo flowing to the deck seal if the cargo tanks have been overfilled.

Deck isolating valve   This isolates the deck main from the rest of the system when the plant is shut down. It also permits maintenance work to be carried out on the non-return valve.

Pressure vacuum breakers   These liquid-filled breakers are fitted in the deck gas main to prevent the cargo tanks from being subjected to an excess positive pressure and to an excess negative pressure.

Deck supply main   This runs forward from the deck isolating valve and delivers the inert gas during operational phases of the voyage.

Branch lines   These pipes deliver the inert gas from the deck main to individual cargo tanks. Another branch line is in the form of a mast riser to allow the venting of gas to the atmosphere.

Basic requirements of inert gas system

The regulations should be studied in full but each system must be capable of:

1 Maintaining the atmosphere of a cargo tank at an oxygen content not exceeding 8 percent by volume and at a positive pressure at all times.

186

2 Delivering inert gas to cargo tanks at a rate of 25 percent more than the maximum discharge rate of the cargo pumps.
3 Delivering inert gas with an oxygen content of not more than 5 percent by volume.

At least two blowers are required to be fitted which together must be capable of delivering the above rate of inert gas to the tanks. Instrumentation must be provided:

1 To indicate continuously the temperature and pressure of the inert gas at the discharge side of the blowers.
2 To indicate continuously and record permanently:
    (a) the pressure of the inert gas supply main forward of the non-return devices
    (b) the oxygen content of the inert gas in the supply main on the discharge side of the blower
3 On the bridge to indicate the inert gas pressure forward of the non-return devices (this should be in the form of a meter).
4 In the machinery control room to indicate the oxygen content of the inert gas on the discharge side of the blower (this should be in the form of a meter).

Audible and visual alarms should include:

1 Low water flow rate to the scrubber.
2 High water level in the scrubber.
3 High gas temperature at the discharge side of the blowers.
4 Failure of a blower.
5 Oxygen content in excess of 8 percent.
6 Failure of the power supply to the gas pressure regulating valve.
7 Low water level in the deck seal.
8 Low gas pressure in the deck supply main.
9 High gas pressure in the deck supply main.

Some of the above alarms also initiate automatic shut-down devices within the system.

All tankers fitted with an inert gas system must also have a closed ullage system.

Basic operating procedure of plant

1 Open and secure the scrubber and water seal overboard discharge valves.
2 Start the water supply systems to the scrubber and deck seal at least 15 minutes before commencing the operation.
3 Ensure that the flue gas is of a suitable quality.

4 Open the flue gas isolating valves.
5 Open the blower suction and delivery valves.
6 Start the blowers.
7 Ensure that the gas regulating valve is open and open the deck main isolating valve.
8 Open the mast riser to allow inert gas to vent to the atmosphere in case it is of poor quality.
9 Check all monitors and when satisfied open the branch line valves to the cargo tanks and close the mast riser valve.

When all the tanks are at the required pressure, shut the deck isolating valve, shut down the blower and close the suction and delivery valves, and close the flue gas isolating valves. The scrubber water should be run for an hour to flush out the tower and drain line, and the blower impellers should be washed with the fresh-water spray to remove grime.

Use of inert gas system during tanker operational cycle

Tank washing on the ballast voyage   When the vessel sailed from the discharge port all the tanks would have been fully inerted and under positive pressure. Tanks may have to be washed during the voyage for the provision of clean ballast and before each tank is washed the oxygen content must be measured at a level of 1 metre below the deck and also at a level half-way down the ullage space. This precaution should be carried out at several locations within each tank. Washing must only commence when the oxygen content of the tank atmosphere is 8 percent or lower. The use of portable atmosphere sampling equipment is described under 'Entry into Enclosed Spaces' in Chapter 1.

The inert gas plant should be started shortly before the tank washing operation commences so that the tanks are under positive pressure. It may be necessary to vent some of the gas to atmosphere via the mast riser to avoid overpressurizing the tanks. On some vessels inert gas will leak to the atmosphere during the washing procedure as it may be necessary to open some tank apertures. Two basic points must be strictly observed on all vessels:

1 The oxygen level of the inert gas being delivered must never exceed 8 percent.
2 The pressure of the atmosphere in the tank must always be positive.

If either of these two basic criteria is not met, tank washing must be stopped immediately. The oxygen content and the gas pressure should be continuously recorded during the washing process.

Changing ballast Before commencing to take in clean ballast to the washed tanks the supply of inert gas to those tanks must be shut off. Arrangements must be made to vent the inert gas in the clean tanks to atmosphere when the ballast water is loaded. This venting is carried out through the inert gas mast riser or by use of an individual tank venting system. As the 'dirty' ballast is pumped out, or transferred to the slop tank, the tanks which have been used for the dirty ballast must be inerted. When the change of ballast operation is completed, riser valves should be shut, and all tanks brought to a slight positive pressure.

At the loading port The inert gas system must be in operation before the pumps start to pump out the ballast and should be left in full operation until the discharge of ballast is completed. The following points should be observed:

1 All cargo tanks should be common with the inert gas main.
2 All mast riser isolating valves should be closed.
3 All monitors and controls must be operating efficiently.

When the ballast has been discharged and all the tanks are fully inerted the system must be shut down and secured before loading commences. The following checks should be made:

1 That the deck isolating valve is closed.
2 That the mast riser venting valves are open.
3 That any stop valves in branch lines are open.

The inflowing cargo will displace the inert gas from the tanks being loaded and the gas will vent to the atmosphere. When loading is completed the riser valves should be closed.

The loaded passage A positive inert gas pressure should be maintained during the loaded voyage to prevent oxygen entering the cargo compartments and to compensate for any loss of inert gas. The plant may therefore be run for a short period every few days to 'top-up' the inert gas in the cargo tanks. The complete system should be checked a few days before entering the discharge port.

The discharge port It may be necessary to depressurize the cargo tanks upon arrival to permit manual ullaging and for water dips to be taken. The tanks must be re-pressurized before discharging commences. The discharge of cargo procedures are similar to those which were observed for the discharge of ballast at the loading port. Some vessels will conduct crude oil washing during the discharge and the same basic safety parameters which were observed during the water tank washing on the ballast voyage must be adhered to. The oxygen content and the pressure of the inert gas in the deck main must be recorded continuously during the discharging operation.

If some cargo tanks require to be ballasted at the end of discharge the standards which were maintained during loading must once again be applied.

Tank washing

Cargo tanks are cleaned for several reasons, the most frequent being the need for clean ballast. When a ship has discharged her cargo she is too light to sail in safety and ballast must be pumped into some cargo tanks. In order to avoid pollution at the loading port, some tanks must be washed during the voyage and filled with sea water; thus we obtain clean ballast. The dirty ballast is then disposed of by using the 'Load on Top' method which will be described later.

Crude oil leaves sediment on tank bottoms which is known as sludge. Washing is the main method to prevent sludge accumulation.

Tanks are also washed when a change of cargo occurs, when tanks need inspecting, or to meet repair work and shipyard preparation requirements.

Washing is carried out by the use of cleaning machines which deliver sea water through two nozzles 180° apart, under high pressure, and in rotating arcs which should cover all of the tank. The water is supplied by a special pump or by the main cargo pumps and the delivery line to the machines has a branch line which is routed via a heater when hot sea water is required. For safety and operational reasons the use of hot sea water should be avoided if possible.

There are two basic types of machine:

Portable machines   These are screwed on to 2.5 inch rubber hoses which are connected to hydrants on the wash water line. The machines are then hung through special openings in the deck. A bonding wire runs through the rubber hose from the coupling at the machine to the hydrant coupling; this provides electrical continuity to prevent static build-up. Portable machines are generally of low capacity from approximately $32m^3$/hour, to high capacity of $75m^3$/hour, depending on machine size and operating pressure (which can be as low as 4.0 kg/cm^2).

Fixed machines   These are similar in construction but connected to permanent supply pipes which 'hang' approximately two metres below the deck and which have a permanent pipeline system above deck used solely for the purposes of the machines. These machines are high-capacity high-velocity and while some portable machines can operate at pressures up to 150 psi (10.5 kg/cm^2) fixed machines often exceed that pressure. A fixed machine

190

operating at a pressure of 10 kg/cm^2 can deliver 150m^3 of liquid each hour through a 38 mm diameter nozzle or 100m^3/hour through a 29 mm diameter nozzle.

Tank washing with water

The general safety precautions stated earlier must be carried out and it is advisable to use a check list or permit-to-work form.

The washing operation can be carried out by using the following method:

1 The plates are removed from the deck openings.
2 The portable machines are connected to the hoses, hung through the openings and the pumps started.
3 The tank is washed in various steps depending upon its size. A typical sequence would be to wash in, say, three stages called drops:
   First drop four metres below the deck. The machines are hung at this level and remain until this part of the tank is thoroughly washed.
   Second drop about halfway down the tank. The hoses are lowered without turning off the water.
   Third drop just above the tank bottom framing. The machines are kept at this level until the bottom is clean.
4 The stripping pump is started at the commencement of the operation and the washings pumped to the slop tank. It is essential that no build-up of water occurs, otherwise the bottom will not be cleaned.
5 When the tank is clean, the pumps are stopped and the hoses hauled on to the deck before disconnection takes place.

The time of the operation depends on several factors, e.g. cargo carried, tank condition, power of machines, water temperature, blind spots, etc. If you are lucky the operation may only take 3–4 hours. However, spot washing may be required to clean 'awkward' corners and machines may be left running at the bottom drop to remove built-up deposits of sludge. Fixed machines work in pre-set cycles of 1–2 hours and can wash a tank efficiently in 3 hours.

A name which is synonymous with tank washing is that of Butterworth who have been manufacturing machines for many years. Figure 6.3 shows a Butterworth portable tank washing machine and Figure 6.4 shows a fixed washing machine.

Tank washing atmospheres

In order to understand this section you must have knowledge of certain definitions.

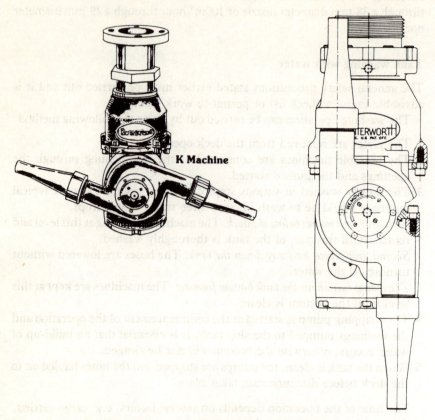

K Machine

Figure 6.3   Portable tank washing machine

Flammable range (also known as explosive range)   Petroleum vapours are only flammable when mixed in a certain volumetric percentage with air. The percentage of the flammable mixture varies with different products. For crude oil the range is 1–10 percent hydrocarbon gas with air, and with kerosene 0.6–6 percent. You must know the flammable range of your cargo.

Lower flammable limit (LFL)   The concentration of a hydrocarbon gas (i.e. gas from petroleum) in air below which there is insufficient gas to support combustion. In non-technical terms the mixture is too lean.

Upper flammable limit (UFL)   The concentration of gas in air above which there is insufficient air to support combustion. The mixture is too rich.

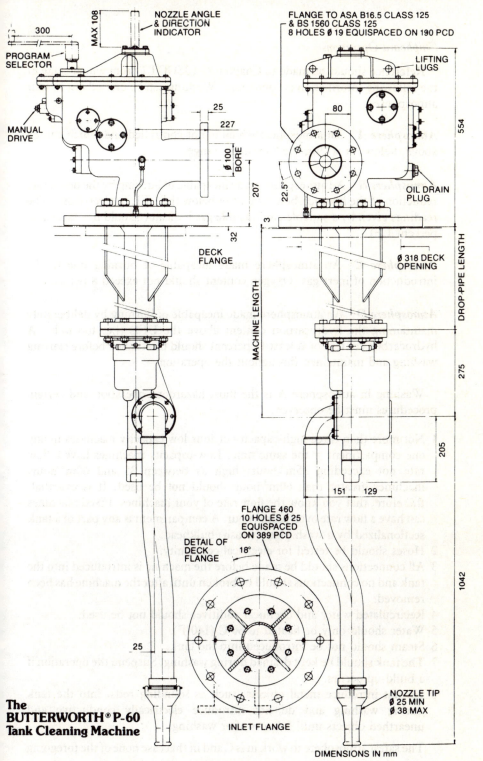

The BUTTERWORTH® P-60 Tank Cleaning Machine

Figure 6.4   Fixed tank washing machine

193

Reference should be made to Chapter 8 of ISGOTT from which some of the following information is condensed. Washing can be carried out in four atmospheres.

Atmosphere A   An atmosphere which is not controlled and which can be above, below, or within the flammable range.

Atmosphere B   An atmosphere made incapable of burning by the deliberate reduction of the hydrocarbon content to below the LFL (i.e. too lean). The readings given by a suitable combustible gas indicator should not exceed 50 percent LFL.

Atmosphere C   An atmosphere made incapable of burning due to the introduction of inert gas. Oxygen content should not exceed 8 percent.

Atmosphere D   An atmosphere made incapable of burning by deliberately maintaining the hydrocarbon content above the UFL (i.e. too rich). A hydrocarbon content of at least 15 percent should be attained before starting washing and maintained throughout the operation.

Washing in atmosphere A is the most hazardous operation and certain procedures must be observed:

1 Not more than three high-capacity or four low-capacity machines in any one compartment at the same time. Low-capacity machines have a flow rate not exceeding 35m^3/hour; high as between 35 and 60m^3/hour; machines greater than 60m^3/hour should not be used. It is essential, therefore, that you know the flow rate of your machines. Fixed machines can have a flow rate over 100m^3/hour. A compartment is any part of a tank sectionalized by a wash plate (swash) bulkhead.
2 Hoses should be tested for electrical continuity.
3 All connections should be made before the machine is introduced into the tank and no connections should be broken until after the machine has been removed.
4 Recirculated water and chemical additives should not be used.
5 Water should only be heated to 60°C (140°F).
6 Steam should not be introduced into the tank.
7 The tank should be kept drained during washing. Suspend the operation if a build-up occurs.
8 Do not introduce metal objects, such as sounding rods, into the tank during washing and do not introduce electrically conductive and unearthed objects until 5 hours after washing.

The safest atmosphere to work in is C and in that case none of the foregoing precautions need be observed. It is the only atmosphere in which the author would be contented to carry out tank washing. However the inert gas plant must be in good condition, the inerting of the tank must be maintained constantly, and the oxygen content must never rise above 8 percent. A positive pressure must be maintained on the system and the gas monitored constantly. Suspend washing if the oxygen content rises above 8 percent.

Refer to Chapter 8 in ISGOTT for washing in atmospheres B and D, and to Chapter 17 for the use and limitations of portable measuring instruments.

Crude oil washing (COW)

Regulations

The IMO International Conference on Tanker Safety and Pollution Prevention which was held in 1978 issued Protocols to existing SOLAS and MARPOL conventions. The MARPOL 1978 Protocol requires COW in new crude carriers of 20,000 ton dwt and above, and existing tankers of 40,000 tons dwt and above must have either COW or SBT (segregated ballast tanks). In accordance with the SOLAS 1978 Protocol every COW tanker must have an inert gas system. The definitions of 'new' and 'existing' are given in the SBT section.

Manuals

Students should refer to *Guidelines for Tankwashing with Crude Oil* published by ICS and OCIMF, and *Crude Oil Washing Systems* published by IMO.

General aspects of COW

Crude oil washing is a process whereby part of the cargo is circulated through the fixed tank cleaning equipment to remove the waxy asphaltic deposits which the cargo has left on the tank. This is normally carried out during discharging. COW has proved to be more effective than water washing for this purpose because the oil acts to disperse and suspend the sediments in the oil and tends to restore the cargo to its as-loaded condition.

Water washing is also necessary if the tank is required for clean ballast or for entry.

Effects of sediment

1 Reduces the effectiveness of drainage.
2 Prolongs the draining operation.
3 Combines with oil, etc. to produce sludge.
4 Impedes the provision of clean ballast.

5 Makes gas freeing difficult and can cause pockets of hydrocarbon gases to form.

6 Reduces the carrying capacity.

7 Reduces ships' earnings.

Advantages of COW

1 Reduced risk of pollution.

2 Reduction in time of passage tank cleaning.

3 Reduction in the cost of tank cleaning (both routine and for dry dock).

4 Reduction in de-sludging costs.

5 Reduction of salt water discharged to the refinery.

6 Reduced corrosion as less salt water is introduced into the tanks during washing.

7 Increases the time available for maintenance.

8 Increased carrying capacity (less slops carried).

9 Increased discharge rates on overall stripping time.

10 Increase in the refinable material discharged.

Disadvantages

1 More crew training required.

2 Increased work load in port.

3 Possible reduced discharge rate on some types of VLCC.

4 Due to the high pressure at which the oil 'jets' strike bulkheads, etc. some structural damage to tank members may be experienced.

Requirements to carry out COW

1 An efficient inert gas system. Crude washing without an IGS is both dangerous and illegal.

2 Fixed tank cleaning installation of a suitable type.

3 An effective monitoring system.

Operating and safety procedures

The use of machines fitted within the tanks avoids the need for use of deck openings and thus prevents the escape of oil or vapour. The machines are provided with the oil via junction lines from the discharge lines of the main cargo pumps.

Before COW is commenced any water in the cargo which may have settled as a water bottom should be drawn from each tank. The slop tank, if it has been used for load-on-top purposes, should be completely discharged and refilled with fresh crude before using it for washing purposes. Blank off any part of the washing system which extends into the engine room and blank off the water heater if one is fitted.

The operation should only be carried out when the tank oxygen level is below 8 percent.

Before use the washing system should be pressure tested for leaks. The system should be checked frequently during the operation, drained and flushed with water after use. A cautionary notice should be posted where applicable: 'The tank washing lines on this ship may contain crude oil. On no account are valves on this line to be opened by unauthorized personnel.'

IMO has laid down specifications for the design of COW systems and operators should refer to the manual, e.g. direct impingement of washing jets is required on 85 percent of the vertical sides of the tank and on 90 percent of horizontal areas within the tank.

There are two basic washing methods. In the 'multi-stage' method the tank sides and structure are washed as the cargo level falls and the bottom is washed as the tank empties. With the 'single-stage' method the tank is first emptied, stripped and then washed in a similar way to the conventional water-washing technique. During each discharge the tanks required for clean ballast should be washed plus 25 percent of the other tanks on a rotational schedule. After both methods pumps and lines should be flushed before ballast is loaded. There will be a thin film of oil on this dirty ballast and IMO regulations require measurement of the surface oil before the ship leaves port. The volume of oil must not exceed an agreed ratio of the cubic capacity of the cargo tanks in which it is contained.

Personnel must be trained for the dual operation of discharging and washing at the same time. Port authorities and terminal operators must be notified of the intention to tank wash and a standard check list must be filled in. All COW operations should be entered in the 'Oil Record Book'.

Gas freeing

This operation normally takes place after water tank washing and all the attendant safety procedures should still be in force. It is standard practice to wash through the whole pipeline system and the pumps; in addition to cleaning the lines this helps to gas free them.

The inert gas system must be shut down and the pressure in the tanks relieved by opening the purge pipes. The tank lids are then opened and the inert gas system blanked off. It is possible to use the inert gas system for gas freeing. However, I have seldom used that method, preferring to use the system only for its important primary function of inerting tanks. The pressure-vacuum valves of all tanks whose vent systems are common with those to be gas freed must now be secured.

Remove any portable washing machines from the tanks and open all deck openings. Air is blown into the tanks through some of the openings by the use of portable high-capacity blowers powered by steam, air or water. The latter are simply water-driven turbine fans which connect via rubber hoses with the wash deck line. Check that the fans are operating at the correct pressure. When going at full blast they produce a high-pitched whine. Extension pieces, shaped like open-ended cylinders, can be inserted into the deck openings before placing the fans. These give a 'jet-like' motion to the air and thus increase the force of the air flow. The fans can also be used as extractors but at decreased efficiency.

The forced ventilation is continued with the tank atmosphere being tested at regular intervals for hydrocarbon gases using an explosimeter, and for oxygen using an analyser. Ventilation is continued until the tank atmosphere is safe, i.e. a reading of 0 percent on the explosimeter and 21 percent on the analyser. The limits which some officers use of 1 percent hydrocarbons and 18 percent oxygen are unsafe. When testing the atmosphere, samples should be taken at the tank bottom and at several depths, using several openings and stopping the ventilation during the tests. Each tank should be gas free in 3–4 hours.

Before entering any tanks use the 'Enclosed Space Procedures'.

Pressure-vacuum valves

Pressure-vacuum valves (PV valves) are used for automatic regulation of pressure within enclosed spaces, particularly cargo tanks. Such valves can be spring-loaded, weighted, or may operate by means of hydraulic action. An easy type to draw and describe is the spring-loaded valve shown in Figure 6.5.

The springs are tensioned so that the valve opens when pressure in a tank rises to about 2 psi (0.137 bars) above atmospheric pressure, and also when it falls to about 0.5 psi (0.034 bars) below atmospheric pressure. A by-pass valve is incorporated into the system to allow unimpeded venting when required, e.g. during loading.

Expansion of the liquid (or evaporation) in the tank will cause the pressure to increase. Once the pressure reaches the pre-set value spring B will be forced to lift, allowing the tank atmosphere to escape up the mast riser and thus decreasing the tank pressure. When the pressure decreases sufficently the plate attached to spring B will re-seat. Contraction of the cargo will decrease the tank pressure. This eventually causes a vacuum but once the pressure falls to 0.5 psi below atmospheric, spring A will lift to allow air into the tank, thus increasing the pressure. When pressure increases sufficiently the plate attached to spring A will re-seat.

An increase of pressure decreases evaporation. This is why the pressure

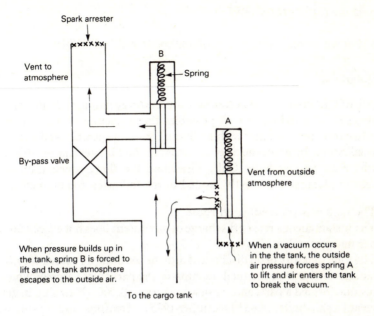

Spark arrester

Vent to
atmosphere

B

Spring

A

By-pass valve

Vent from outside
atmosphere

When pressure builds up in
the tank, spring B is forced to
lift and the tank atmosphere
escapes to the outside air.

When a vacuum occurs
in the the tank, the outside
air pressure forces spring A
to lift and air enters the tank
to break the vacuum.

To the cargo tank

Figure 6.5   Pressure-vacuum valve

side of the valve is set to 2 psi. Evaporation could be completely prevented if
the tanks were sufficiently pressurized. However, conventional cargo tanks
are only strong enough to withstand 3.5 psi (0.24 bars) above atmospheric
pressure and for reasons of safety a pressure of 2.5 psi (0.17 bars) should
never be exceeded.

PV valves are sometimes arranged in tank groups and it is important to
fully understand the grouping system in order to avoid accidental
contamination when grades are being loaded and carried, through
hydrocarbon gases being fed to tanks along common vent lines. For this
reason the position of block valves should be known. Some vessels have a
'self-contained' venting system for each tank, therefore as a tank is loaded the
PV valve for that tank should be re-set to the normal position. On ships with a
group system the valves can only be re-set when each group of tanks has been
loaded.

PV valves require careful and frequent maintenance and should be opened
at regular intervals to ensure that they are not sticking, or rusting too much,
and that the seats are in good condition. The spark arrestor gauze must be
kept clean and renewed frequently. During loaded passages a check should
be made that each valve is working.

'Load on top' system (LOT)

Also known as the 'retention of oil on board process' (ROB).

Regulations

The IMO International Convention for the Prevention of Pollution from Ships 1973 (MARPOL 1973) prohibits the discharge of any oil or oily mixture from an oil tanker within 50 miles from the nearest land and within Special Areas. In brief, the Special Areas are the Mediterranean, Black Sea, Baltic, Red Sea, and the Gulf. Elsewhere the flow, concentration, and quantity discharged is strictly limited to the following requirements:

1 The tanker is proceeding en route.
2 The instantaneous rate of discharge of oil content does not exceed 60 litres per nautical mile.
3 The total quantity of oil discharged into the sea does not exceed for existing tankers 1/15,000 of the total quantity of the particular cargo of which the residue formed a part, and for new tankers 1/30,000. (New ship in general means a ship built after 31 December 1975 and existing ship is a ship which is not new.)
4 The tanker has in operation an oil discharge monitoring and control system and a slop tank arrangement.

Manuals

Students should refer to the above Convention, plus *Clean Seas Guide for Oil Tankers* and *Monitoring of Retained Oil Residues* published by ICS and OCIMF.

Operation

The LOT system meets the above regulations by using the principle of decantation to separate oil residues from water. The sea water is then discharged at sea while the oil residues are retained on board. Thus the tanker enters the loading port with clean ballast, the residues are co-mingled with the new cargo, and pollution is avoided. The operation can be divided into various stages:

1 Dirty ballast is loaded into certain tanks on the completion of cargo discharge.
2 During the voyage the tanks to be filled with clean ballast are washed, the oil water mixture being stripped to the slop tank.
3 The clean tanks are filled with ballast.

200

4 During this period the oil in the dirty ballast has floated to the top. If the ship has carried out crude oil washing, the oil skim is already on the surface. The clean water under the oil is discharged and the oil plus some water is transferred to the slop tank.

5 The mixture in the slop tank is allowed to settle for at least 24 hours. If voyage requirements permit, a period of several days is allowed for this process.

6 The clean water under the oil is pumped out to sea, being constantly monitored to stay within the regulations.

7 The next cargo is loaded on top of the oil residue and the mixture is pumped ashore at the discharge terminal.

It should be noted that the slop tank decanting process cannot be carried out within 50 miles of the coast or within Special Areas. Ensure that the 'Oil Record Book' is correctly kept and that a 'ROB monitoring record' is maintained.

Segregated ballast tanks (SBT)

Regulations

MARPOL 1973 (Regulation 13) stated that every new oil tanker of 70,000 tons deadweight and above should be provided with segregated ballast tanks. The broad definition of segregated ballast means ballast water introduced into a tank which is completely separated from the cargo oil and which is permanently allocated to the carriage of ballast. The capacity of such tanks should be so as to allow the ship to operate safely on ballast voyages. However, if the Master deems it necessary in extreme weather conditions, additional ballast may be carried in oil tanks to ensure the safety of the ship. Such additional ballast must comply with LOT procedures.

The 1978 MARPOL Protocol extended the above requirements:

1 All new crude oil tankers of 20,000 dwt and above must be provided with SBT and operate COW procedures.

2 All new product carriers of 30,000 dwt and above must be provided with SBT.

3 All existing crude oil tankers of 40,000 dwt and above must either be provided with SBT or operate COW procedures.

4 All existing product carriers of 40,000 dwt and above must be provided with SBT.

There are acceptable alternatives to SBT but if these are required, Regulation 13A of MARPOL 1978 should be studied in full.

New ship in general means a ship for which a contract was placed after 1 June 1979, or was in an early stage of building on 1 January 1980, or was delivered after 1 June 1982. Existing ships refers to ships which were in existence on the date of entry into force of the Protocol.

The date of entry into force of MARPOL 73/78, i.e. both MARPOL 73 and the Protocol of 78, was 2 October 1983.

Further reading

BAPTIST, G. *Tanker Handbook for Deck Officers* (Brown: Glasgow, 1975).
CORKHILL, M. *Product Tankers and their Market Role* (Fairplay: London, 1978).
KING, G. A. B. *Tanker Practice* (Maritime: London, 1969).
MARTON, G. S. *Tanker Operations* (Cornell: Cambridge, Maryland, 1978).
RUTHERFORD, D. *Tanker Cargo Handling* (Griffin: London, 1980).

ICS Publications

Ship to Ship Transfer Guide (Petroleum).
Prevention of Oil Spillages through Cargo Pump Room Sea Valves.

Journal

BRIDGES, F. V. 'Operational oil spillages', *Seaways*, April 1983, pp. 13–17.

7

Cargoes Which Are Intrinsically Dangerous

A Dangerous Goods
B Liquefied Gas Cargoes
C Chemical Cargoes

A Dangerous Goods

Merchant Shipping (Dangerous Goods) Regulations 1981

These regulations give effect to Chapter VII (Carriage of Dangerous Goods) of SOLAS 74.

The shipowner or employer is required to make arrangements for ensuring the health and safety of persons working on board in operations dealing with dangerous goods. The term 'dangerous goods' refers to goods which are classified in the 'Blue Book', the 'IMDG Code', and in certain IMO publications such as the 'Bulk Dangerous Chemicals Code', and to any other substance which might be dangerous if transported by sea.

The regulations apply to United Kingdom ships and to other ships operating in United Kingdom waters.

Documentation of packaged dangerous goods

No dangerous goods shall be loaded unless the shipper has provided a dangerous goods declaration; not to do so is an offence. The declaration must give the correct technical name of the goods (i.e. a description of the goods which readily identifies their dangerous characteristics, including any shipping name given in the IMDG Code), the identity of the goods, the United Nations number if applicable, and must also indicate the class in which the goods belong.

Class 1   Explosives

Class 2   Gases compressed, liquefied or dissolved under pressure, subdivided into three categories:

    2.1 Flammable gases

    2.2 Non-flammable gases, being compressed, liquefied or dissolved, but neither flammable nor poisonous

    2.3 Poisonous gases

Class 3   Flammable liquids, subdivided into three categories:

    3.1 Low flashpoint

    3.2 Intermediate flashpoint

    3.3 High flashpoint

Class 4.1 Flammable solids

Class 4.2 Substances liable to spontaneous combustion

Class 4.3 Substances which in contact with water emit flammable gases

Class 5.1 Oxidizing substances (agents)

Class 5.2 Organic peroxides

Class 6.1 Poisonous (toxic) substances

Class 6.2 Infectious substances

Class 7   Radioactive substances

Class 8   Corrosives

Class 9   Miscellaneous dangerous substances which present a danger not covered by other classes

In addition to the declaration, the shipper must supply the following written information where appropriate:

(a) the number and type of packages

(b) the gross weight of the consignment

(c) the net weight of the explosive content of Class 1 goods

(d) the flashpoint if 61°C or below.

If goods are packed into a container or vehicle the vessel must be given a packing certificate for the container or vehicle.

A stowage plan should be made which gives the information noted above and also the location of where the goods are stowed.

Packaging of dangerous goods

Goods should be packed in such a manner that, having regard to the properties of the goods, they withstand the ordinary risks of handling and transport by sea. The goods must also be packed in accordance with the requirements of the Blue Book or, if there is no such requirement, then in accordance with the IMDG Code. Any portable tank or road tank vehicle which contains dangerous goods must be certified under the provisions of Annex 1 or 2 of the Blue Book. The shipper must provide a declaration which states that the goods are packed in accordance with the regulations.

Marking

The following requirements should be complied with:

1 The package must be clearly marked with the correct technical name of the goods and an indication should be given as to the danger which could arise during the transportation of the goods.
2 The markings must comply with the Blue Book or IMDG Code.
3 If the outer material of the package will survive three months' immersion the marking must be durable.
4 If the outer material will not survive three months any inner receptacles which will survive three months must be durably marked.
5 If the goods are carried in a container or similar unit, then that unit must have distinctive labels on the exterior which comply with the Blue Book or the IMDG Code class label system. In the case of a container or tank the labels should be on each side and end. If a vehicle, the labels should be on each side and at the rear.

Stowage and packing

Dangerous goods should be stowed in a manner which is safe and proper and in a location where the appropriate adequate ventilation can be provided. Goods in a container or similar unit should be packed in a safe and proper manner. Goods which are liable to interact dangerously must be effectively segregated from one another.

Carriage of explosives other than safety explosives

Explosives should only be carried in a compartment in which any electrical apparatus and cables are designed and used so as to minimize the risk of fire or explosion.

Where Category II stowage is required by the IMDG Code, the explosives should be stowed in a magazine which shall be kept securely closed when the ship is at sea.

Detonators must be effectively segregated from all other explosives.

Carriage of dangerous goods on passenger ships

No explosives can be transported on a ship carrying more than 12 passengers except:
(a) safety explosives
(b) any explosives the net weight of which is 10 kg or under
(c) distress signals up to a total weight of 1000 kg
(d) fireworks which are unlikely to explode violently, provided that no fireworks are carried on ships transporting unberthed passengers

205

The stowage of distress signals and fireworks must be supervised.

No dangerous goods should be allowed on board any vessel carrying more than 25 passengers.

Dangerous goods loaded in bulk

Bulk dangerous goods include solid bulk cargoes as well as liquid chemicals and gases. Notification of such goods must be given to the owner or master and the goods will be deemed to be loaded acceptably if the relevant IMO Code is adhered to.

Blue Book

This is a Department of Transport publication entitled *Carriage of Dangerous Goods in Ships* and which in general is a reflection of the IMDG Code. The 1984 edition is constantly updated by amendments which should be inserted into the appropriate section. The Blue Book is the report of the Standing Advisory Committee on the Carriage of Dangerous Goods in Ships and the Committee believes that safety at sea 'will be facilitated if the government can accept virtually all the recommendations of the IMDG Code'. The Committee therefore recommends that the United Kingdom adopts the IMDG Code in 'large measure' while recognizing that for a period of time there will be 'certain minimal variations' between the Blue Book and the IMDG Code.

From the above it would be reasonable for British mariners to consider both publications as companion and complementary codes.

The Blue Book is a loose-leaf one-volume publication which contains general and specific recommendations for the safe transportation of dangerous goods. Mariners should always consult the Blue Book when in any doubt as to the hazardous nature of any cargo to be loaded.

IMDG Code

The *International Maritime Dangerous Goods Code* is published by IMO in five volumes. The Code lays down certain basic principles concerning the transportation of dangerous goods. Recommendations for good practice are included in the classes dealing with such substances, and detailed recommendations are given for individual substances. The code is published in a loose-leaf format in five volumes. Amendments are published regularly to keep pace with the technological developments of cargoes. Volume 1 contains a general introduction which gives detailed information on the application of the Code, classification, documentation and labelling, packing and packaging, stowage and segregation, fire prevention and fire fighting, and other related subjects. Substances are grouped together under the various class headings. Each class has an introduction giving information on properties, packing, stowage, segregation, fire fighting, and limited quantities. Each substance usually has a page (schedule) to itself and the information includes the chemical name and formula, the United Nations number, properties, packing, stowage, and observations. A general index of technical information is included in Volume 5, and the index should always be consulted when attempting to locate the appropriate schedule for any substance or article. The IMDG stowage and segregation procedures and requirements should be carefully studied before a decision is taken as to the actual stowage position of dangerous goods.

Emergency Procedures for Ships Carrying Dangerous Goods

This is an IMO publication which gives information concerning the safety, first aid, and emergency procedures to be followed and action to be taken in the event of an incident involving certain dangerous goods. The goods are classified in accordance with Chapter VII of SOLAS and the recommended procedures should be used in conjunction with the information provided in the IMDG Code and the IMO publication *Medical First Aid Guide for Use in Accidents involving Dangerous Goods* (MFAG).

The Emergency Schedules (EmS) are divided into five sections:

1 Group title with the emergency schedule number (EmS No.).
2 Special equipment required.
3 Emergency procedures.
4 Emergency action.
5 First aid

The appropriate schedules should be consulted before goods are loaded in order to ascertain that the vessel has the correct equipment to deal with any incidents which might occur. The schedules are also useful when planning emergency team training exercises.

M Notices

New amendments to the Blue Book and IMDG Code are promulgated in M notices. The notices referring to dangerous goods are too numerous and varied to be listed here. A full list of current Merchant Shipping Notices should be consulted but M1117 and M1124 must be read.

M1213 draws attention to The Merchant Shipping (Dangerous Goods) (Amendment) Regulations 1986, especially to Regulation 2(6) which 'requires the owners and masters of, as well as employers of persons aboard, ships in which dangerous substances are being carried, stowed, handled, loaded or unloaded to ensure that, so far as is reasonably practicable, nothing in the manner in which those substances being carried, stowed, handled, loaded or unloaded is such as might create a risk to the health and safety of any person'. The purpose of the Notice is to give practical guidance on the implementation of Regulation 2(6) and to this end an Annex is attached to the Notice which contains a Code of Practice based on IMO recommendations on the safe transport, handling and storage of dangerous substances in port areas.

The Annex is comprehensive but it particularly draws the attention of masters to ensuring that 'the loading or unloading of dangerous substances is carried out under the overall supervision of a duly qualified officer of the vessel and that the officer is aware of the risks involved, the correct procedures to be followed, and the steps to be taken in an emergency'. The master should also, as far as is reasonably practicable, ensure that no intoxicated person has anything to do with any aspects of such operations. The master or owner should ensure that information on the dangerous substance and on the stability of the vessel is in the possession of the duly qualified officer and that the information is available for the use of the emergency services.

The 'Code of Practice for the Handling, Loading and Unloading of Dangerous Goods in Harbour Areas', which comprises most of the Annex, also deals with such topics as:

> Transfer of dangerous liquids in bulk
> Special precautions for the handling of explosives
> Precautions against fire and explosions

The Notice also draws attention to 'The Dangerous Substances in Harbour Area Regulations', known as the DSHA Regulations, which became operative on 1 June 1987. These Regulations take IMO recommendations into account (such as the IMO publication *Safe Transport, Handling and Storage of Dangerous Substances in Port Areas*) and were drawn up by the Health and Safety Commission (HSC). The Regulations are enforced by the Health and Safety Executive (HSE) which is the 'operational wing' of the HSC and, as is the case with the new Docks Regulations, these particular Regulations come in the form of a 'package' of Regulations, Schedules, a Guide and an Approved Code of Practice. The Regulations cover the following topics:

> Entry of dangerous substances into harbour areas
> Marking and navigation of vessels
> Safe handling of dangerous substances
> Loading and unloading of dangerous substances
> Liquid dangerous substances in bulk

Packaging and labelling
Emergency arrangements and unusual incidents
Storage of dangerous substances
Handling of explosives

General fire precautions when carrying dangerous goods

The following recommendations are to be found in either the Blue Book or the IMDG Code:

1 Reject any damaged or leaking packages.
2 Packages should be stowed in a location which ensures protection from accidental damage or heating.
3 Combustible material must be kept away from ignition sources.
4 Goods must be segregated from substances liable to start or to spread fires.
5 It may be necessary to ensure accessibility of dangerous goods so that packages in the vicinity of a fire may be protected or moved to safety.
6 Enforce prohibition of smoking in dangerous areas.
7 Post 'No Smoking' signs or symbols.
8 All electrical fittings and cables must be in good condition and safeguarded against short circuits and sparking.
9 All ventilators must have spark arrestors of suitable wire mesh.

Officers must check the individual entries in the codes to ascertain the best extinguishing agent for each particular consignment. In general the most effective agent is water but in some cases the use of water can intensify the fire or cause an explosion. Water should not be used against fires involving explosives. Sprinkling or flooding is the only sure way to deal with temperature rises which may lead to explosive decomposition. Sufficient extinguishers of an appropriate type should be carried.

Some substances on fire may emit poisonous fumes. Therefore, protective clothing and sets of self-contained breathing apparatus should be readily available. The Blue Book recommends that a minimum of two sets should be carried in addition to the SOLAS requirements for general cargo vessels.

It is the general practice to close openings when fighting a fire. However, with substances that either evolve oxygen on being heated or are liable to self-sustaining combustion, such a procedure could prove to be ineffective or dangerous.

General precautions during loading and unloading of Class 1 goods

1 Fire-fighting appliances ready for instant use.
2 Unauthorized personnel cleared of area.

3 'No Smoking' notices posted and enforce prohibition of smoking.

4 Fire detection system tested.

5 No matches or lighters to be carried by anyone involved in the operation.

6 If possible the cargo operations should be carried out during daylight hours, 'B' flag hoisted.

7 Bilges should be cleaned before loading commences.

8 Inspect the ventilation fans to ensure that they are not unsafe.

9 Check the lightning conductors.

10 Only approved lighting to be used.

11 No wireless transmissions to be permitted.

12 The radar should not be operated, the duty officer should have the radar fuses in his possession.

13 The funnel exhaust spark arresters should be in good condition.

14 No bunkering or repair work during the operation.

15 Defective packages should not be accepted for shipment.

16 Fork lift trucks should not be used.

17 Operations should be suspended in rain.

18 All explosives should be tallied.

19 A responsible person should be present during the operation and any magazines should be kept locked when not in use.

General stowage precautions for all Class 1 goods

1 Do not stow in the same compartment with goods liable to give off flammable vapours.

2 Stow in a cool place away from engine room bulkheads and away from the ship's side if sailing to the tropics.

3 Stow away from living quarters.

4 The space should be dry and well ventilated as explosives are unstable when wet.

5 Stow in a convenient location for jettisoning.

6 When possible stow close to the hatch square. If an explosion occurs, the damage may be limited to the hatch cover, leaving the shell plating intact.

7 Check that the fire-fighting system is suitable for dealing with the type of explosives carried.

8 All ventilators should have spark arresters consisting of two layers of fine mesh gauze.

9 If possible, electric cables should not pass through a compartment containing Class 1 goods. The Blue Book should be consulted for exceptions to this recommendation.

10 Artificial lighting used for cargo operations should be of an approved type.

210

11 Electrical fittings in compartments in which explosives are to be stowed should be disconnected.

12 Electrically powered ventilation fans should be flame-proof. If the fans are not flame-proof, they should be disconnected.

13 Detonators must be effectively segregated from all other explosives.

Fixed fire-fighting and detection systems

The existence of several sets of fire regulations can be somewhat confusing so M1117 (Fire protection in ships carrying explosives) should be consulted as it prescribes a 'minimum standard of fire protection considered necessary when UK ships load explosives anywhere in the world and when foreign ships load explosives within the UK'. Ships to which the 1984 Fire Protection Regulations apply will require an automatic fire detection system but it is now recognized internationally that all ships carrying dangerous goods should be protected by a fire detection system in the cargo spaces. The Department of Transport, however, has accepted that on 'pre-1984' ships the following arrangements as being equivalent to a fire detection system on ships carrying explosives for one voyage only:

> Efficient means in the form of pipes, or ventilators if they are of convenient height and position, by which fire in the compartments concerned may readily be detected remotely by sense of smell, combined with regular tours of inspection by a crew patrol to the areas concerned at half-hourly intervals, such patrols to be recorded in the ship's log.

Ships which carry explosives on subsequent voyages must fit a fire detection system.

All ships will also have a fixed fire fighting system.

Stowage categories of Class 1 goods

Different explosives, depending on their properties, require different stowage arrangements. Each different method of stowage is defined as 'Stowage Category'. The stowage categories of goods are indicated on the individual schedules in the IMDG Code.

Stowage Category I is also known as 'ordinary' stowage as no magazine is required. The goods should be stowed in accordance with the general stowage precautions listed above.

Stowage Category II refers to Class 1 goods that require magazine stowage and has been subdivided by adding the letters A, B, or C. The letters indicate a requirement for three different types of magazine. If a Chief Officer finds it necessary to construct a magazine it will usually be Type A, which must be constructed with clean, unimpregnated timber using galvanized iron or non-ferrous metal nails. The following details are a synopsis of the construction requirements and the Blue Book and IMDG Code must be consulted on board ship before constructing such a magazine.

Construction of a Type A magazine

1 The magazine should be a space of the required size, normally in the 'tween deck. Constructed of 25 mm close-fitting boards secured internally to 75 × 75 mm uprights spaced 600 mm apart running from deck to deck and firmly secured top and bottom. When the height of the deck exceeds 2.4 metres the uprights should not be spaced more than 450 mm apart. On steel decks the heels of the uprights should be stepped on, and secured to, a board 25 mm thick laid on the deck.

2 The boards should be securely fastened by 75 mm nails, three to each board on each upright.

3 25 mm thick boards should be fastened to the outside of the uprights at the upper and lower ends for securing the heads of shores when required. The space formed between the lower board and the partition boards should be filled in.

4 The ship's sides and bulkheads may be used as the sides of the magazine but they must be sheathed with wood (for dimensions consult the Blue Book).

5 The magazine flooring should consist of close-fitting 25 mm boards secured to 75 × 50 mm bearers spaced 450 mm apart.

6 All pillars, stanchions, ventilator shafts, etc. must be sheathed.

7 The deckhead need not be sheathed but it must be clean and rust-free.

8 One or more doorways, at least 1.2 metres wide, should be fitted facing the hatchway. The door must have an internal wide 'rabbet'. (Consult the Blue Book for door dimensions.)

9 The door may be made in one or two parts. The lower part to be 1 metre in height and shipped from the inside, upper part to be of such a height as is necessary and to be shipped from the outside.

10 The doors should be fitted with hand holes for lifting purposes.

11 A 75 × 75 mm stanchion should be wedged securely between the middle of the door and the deck.

12 Where insulated spaces are used as magazines, hinged insulated doors should be used.

Type B magazine

As Type A but ship's side, bulkheads, etc., need not be sheathed. The flooring should consist of portable sparred gratings.

Type C magazine

As Type B except that it should be placed at least 2.5 metres from the ship's side.

B Liquefied Gas Cargoes

IMO Codes

The construction and equipment of vessels transporting liquefied gases in bulk is governed by the following codes:

Code for Existing Ships Carrying Liquefied Gases in Bulk	Ships built before 1977
Code for the Construction and Equipment of Ships Carrying Liquefied Gases in Bulk	Ships built after 1976 but before July 1986
International Code for the Construction and Equipment of Ships Carrying Liquefied Gases in Bulk (The International Gas Carrier Code—IGC Code)	Ships built after June 1986

Amendments to the codes are published when technological changes or advancements necessitate some alterations to the requirements. No reference should be made to the codes without checking that the relevant amendments have been correlated with the particular code.

It is not possible in a book of this nature to study the details of the above codes. However, in the general preamble to one of the codes IMO states that its purpose 'is to recommend suitable design criteria, construction standards and other safety measures for ships transporting liquefied gases and certain other substances in bulk so as to minimize the risk to the ship, its crew and to the environment'. One must accept that the possibilities of a gas carrier being involved in a collision or grounding do exist and ships built to the construction code must be able to survive 'the normal effects of flooding following assumed hull damage' which is caused by external force. In addition, the cargo tanks must be protected from penetration damage caused by minor incidents such as striking a jetty and must be given a measure of protection from damage in the case of major incidents such as collision and stranding. This protection is afforded by locating the cargo tanks at specified minimum distances inboard from the hull. The amount of damage which a

213

liquefied gas tanker can endure and still contain the cargo depends upon the type of cargo being carried and the size of the vessel. Those vessels transporting liquefied gases which present the greatest hazards must comply with specifications for a Type IG ship. Vessels carrying cargoes of a decreasingly hazardous nature must be built to the specifications for Types IIG, IIPG, and IIIG accordingly.

There are two basic types of cargo containment tanks, self supporting and non-self supporting. Self supporting tanks (also known as independent tanks) do not form part of the ship's hull and are built with sufficient inherent strength to support the tank structure and contain the cargo (Figure 7.1).

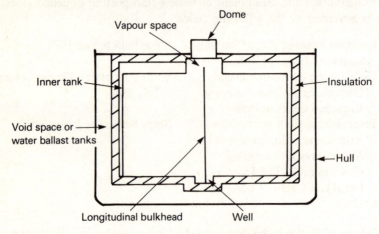

Figure 7.1   Self supporting tank, liquefied gas carrier

Non-self supporting tanks (membrane and semi-membrane) have no inherent strength and are supported by attachments to the insulation; loads are transmitted to the inner hull through the insulation.

The codes also contain stringent requirements for cargo controls, ventilation, instrumentation, fire protection, and operational procedures.

British Regulations

M1237 draws attention to the Merchant Shipping Gas Carrier Regulations 1986, operative 1 July 1986, which make the IGC Code mandatory for UK ships throughout the world and for foreign ships in UK waters. M1170 draws attention to the survey requirements for UK gas carriers and notes that, in addition to an initial survey for the issue of the 'Certificate of Fitness' and a periodical survey for the renewal of the Certificate, annual surveys and at least one intermediate survey are also required during the validity of the Certificate (the annual survey should be held within three months before or after the anniversary date of the Certificate and the intermediate survey within six months before or after the half-way date). The Fitness Certificate lasts for five years and M1170 should be studied for the details of the surveys.

Tanker Safety Guide (Liquefied Gas)

This guide is published by the ICS and its purpose is to provide those serving on gas carriers with information on recognized good operational practices and safe procedures.

It is a loose-leaf one-volume guide which covers all aspects of the operation of vessels carrying gases such as liquefied petroleum gas (LPG), ethylene, liquefied natural gas (LNG), and ammonia. In particular the contents include the following:

1 Information on the general properties of liquefied gases, the hazards of such, and how to transport them in safety.
2 General precautions for the safety of the ship, especially the avoidance of fire.
3 Cargo operational guidance, including loading, discharging, reliquefaction, purging, and sampling.
4 Information on cargo equipment and instrument operation and maintenance.
5 Correct procedures for entry into enclosed spaces, fire fighting, emergency action, and personnel protection.

The guide is written in practical terms, the physical properties of gases are explained at some length and explanations are given as to why special precautions are necessary.

The ICS also publishes a booklet, *Safety in Liquefied Gas Tankers*, and all Department of Transport students should study carefully the contents. The booklet is particularly useful for explaining cargo hazards to crew members.

Some common liquefied gas cargoes

Liquefied natural gas

LNG is a mixture of hydrocarbon gases in a natural state. Methane predominates but rich natural gases will contain percentages of ethane, butane, and propane. The mixture varies from place to place and in particular whether it is associated with crude oil or found in gas pockets.

215

Liquefied petroleum gas

LPG is generally a by-product gas produced by the refining process of crude oil, the main gases being propane and butane.

Chemical feedstocks

These include propylene, butylene, butadiene, and ethylene. The total tonnage of chemical gases transported each year is relatively small but the cargoes have great commercial value and are therefore significant in terms of 'earning power'.

Ammonia

Ammonia is mainly used as a fertilizer. The product is usually shipped from developed countries.

The critical temperature of a gas is the temperature above which a gas cannot be liquefied by pressure alone. Apart from LNG and ethylene, all of the above cargoes are below their critical temperature at ambient temperatures. They can therefore be liquefied by pressure and carried at ambient temperatures. This is not possible for ethylene and LNG because the critical temperatures and pressures are too high. The only practical method of transporting LNG and ethylene is to liquefy the gases by cooling and to carry them in semi- or fully-refrigerated compartments.

Types of gas carriers

There are five basic types of gas carriers.

Fully pressurized

The most 'basic' type of gas carrier. The cargo, which is usually LPG or ammonia, is transported in cylindrical pressure tanks which are designed to withstand pressures of about 15 to 17 atmospheres. The ships seldom have a cargo capacity of more than 1000 m^3 and are engaged in short sea trades.

Semi-pressurized/semi-refrigerated

These ships have tanks which are suitable for carrying cargoes in a temperature range of approximately −33°C to −50°C at a pressure range between 4 and 7 atmospheres. The cargoes include LPG and chemical feedstocks and the ships usually have a cargo capacity up to 12,000 m^3. The ships are fitted with reliquefaction equipment to condense the vapours given off by the liquefied gas during transit.

Fully refrigerated LPG

These ships carry LPG at temperatures between $-47°C$ and $-55°C$ at about atmospheric pressure. The cargo capacity is usually between 5,000 m^3 and 100,000 m^3. Reliquefaction equipment and inert gas plants are fitted.

Ethylene ships

These are specialized ships of 1,000 m^3 to 12,000 m^3 capacity in which the cargo is carried at temperatures around $-104°C$ at about atmospheric pressure. Reliquefaction equipment and an IG plant are both usually installed.

LNG ships

The cargo is fully refrigerated at approximately $-163°C$ and carried at about atmospheric pressure. LNG ships are generally large with a cargo capacity from 40,000 m^3 to 140,000 m^3. An IG plant is installed but due to the low temperatures at which LNG is carried, no reliquefaction equipment is fitted. The cargo vapour is either vented or consumed in the main engine.

Cargo-handling equipment

The cargo-handling equipment is similar to that of oil tankers. Shore cargo arms are connected to the ship's deck-piping system at the manifolds located close to midships. Deck mains run fore and aft from the manifolds and branch pipes lead from the deck mains to the individual tanks. However, there are no fore and aft cargo mains running through bulkheads and along the tank bottoms to a central pump room as on a conventional tanker. Because of the stringent constructional and safety requirements for gas carriers, each tank is an individual unit with its own pumps and vertical cargo branch lines which connect with the deck main.

Cargo pumps

There are two basic types of pump, (a) deep well and (b) fully immersed.

Deep well In this type the electric drive unit is situated on deck and an extended drive shaft inside the discharge pipe leads to an impeller in the suction unit located in a well at the bottom of the tank. This pump has major disadvantages caused by differential expansion of tank, shafting, and piping and is not as popular as the fully immersed pump.

Fully immersed This is situated at the bottom of the tank and is a fully submerged, electrically driven centrifugal pump. The pump sucks the liquid through its centre and the liquid is sent up to the deck via the pump discharge pipe.

Each cargo tank has two pumps. A transferable pump which is used as an emergency cargo pump is also carried.

Reliquefaction plant

During the loaded passage gas will evaporate from the cargo, the process being known as 'boil off'. On LNG carriers the boil off is vented to the atmosphere or, alternatively, is used as propulsion fuel in the boilers. On LPG carriers the gas is withdrawn from the vapour space at the top of the tank and is led via the vapour line to the reliquefaction plant. The purpose of the plant is:
(a) to prevent loss of cargo due to evaporation
(b) to reliquefy cargo vapours produced by the loading process
(c) to maintain cargo temperature and pressure on passage
(d) to cool tanks and pipes before loading
  The plant can be thought of as a large domestic refrigerator within which the vapour is cooled, compressed, condensed, and then returned to the tank via the condensate return line. LPG carriers generally have three storage vessels or tanks which are situated on deck and which contain liquefied gas. This liquefied gas can be used for cooling, purging, and change of cargo operations and the three separate tanks usually contain quantities of butane, propane, and ammonia. Cargo condensate can be led to the deck storage vessels from the reliquefaction plant to be retained for future operations.

Spray rails

These are pipes located within the cargo tanks and used for evenly cooling the tank before loading and also for returning the condensate during passage. The cooling pipes, which can be located at various heights within the tank, contain small holes which produce a fine spray of small liquid droplets. The holes are directed upwards so that the droplets first move upwards before falling, thus prolonging the time available for evaporation. The spray rails used for the returning condensate have larger holes which permit the returning liquid to mix with the cargo before evaporation takes place.

Purge lines

These are used to change the tank atmosphere. Purge lines lead to both the top and the bottom of a tank as the replacing element may be lighter or heavier than the previous atmosphere.

Puddle heaters

Puddle heaters are heating coils situated in the bottom of tanks which are used to vaporize any small amounts of cargo left in the tanks. This procedure is usually one carried out when changing cargoes or to prepare for tank entry.

Relief valves

Relief valves are fitted to cargo tanks so that if tank pressure exceeds the pre-set limits the vapour is vented to the atmosphere. This venting is undesirable as the gas is flammable, toxic, and the cargo loss is costly. Excess tank pressure can be caused by:
(a) cargo not being cool enough during loading
(b) excessive pressure on the vapour return line during discharging
(c) a breakdown in the reliquefaction plant
   Vacuum relief valves are unnecessary as the tanks are always at positive pressure.

Cargo valves

These are operated hydraulically or pneumatically. An automatic quick-closing system can close all cargo-related valves in an emergency.

Equalizing valves

Some cargo tanks are subdivided by a longitudinal bulkhead to reduce free surface effect. A valve, known as an equalizing valve, is located at the bottom of such a bulkhead to:
(a) enable both compartments of a tank to be discharged in the event of a breakdown of the pump in one compartment
(b) keep the liquid at the same level on both sides of the bulkhead during cargo operations in order to prevent a list occurring and to prevent excessive cargo pressure on one side of the bulkhead

Ballast tanks

The cargo tanks are never used for ballast and thus the ballasting arrangements are completely separate from the cargo system. Ballast is carried in upper and lower hopper tanks, wing tanks, and in double-bottoms.

Inert gas plant

Inert gas is required for various cargo operations and the plant is similar to that found on conventional tankers. However, an additional feature on gas carriers is the dessicant drying system which ensures that the moisture content of the gas is kept to an absolute minimum. The nitrogen content of the inert gas is approximately 85 percent. The impurities in inert gas may react with certain chemical gases and therefore inert gas cannot be used for purging those gases, e.g. ammonia reacts with the carbon dioxide in inert gas to produce a carbonate in the form of a white crystalline deposit.

Tank dome

This provides the entry point into the tank for cargo lines, vapour lines, sensors, control and monitoring equipment. The dome is free-floating within a deck aperture and a neoprene gasket, known as the deck seal, ensures gas tightness. Within the dome is the area known as the vapour space. The cargo in a tank never occupies more than 98 percent of the space available for cargo, i.e. the liquid level must always be below the bottom of the vapour space. The entry to the vapour suction line must always be well above the cargo level as any liquid entering the vapour line would damage the compressor in the reliquefaction plant.

Tank monitors and controls

There are three basic monitoring parameters:
(a) cargo and tank temperatures
(b) cargo liquid levels
(c) tank pressure
The quality of the inert gas must also be monitored constantly.

In addition to the monitoring function, tank instruments also activate alarms and in some cases automatically control equipment such as cargo valves.

(a) Temperature

Probes containing sensors are located at various positions throughout a cargo tank to monitor the cargo and the tank structure temperatures. It is essential that the temperature gradient between the bottom and the top of the tank does not become excessive, otherwise unacceptable stresses will be set up.

(b) Liquid levels

The level of the cargo can be measured by a variety of devices, e.g. float gauges, pressure gauges, and ultrasonic gauges. As it is extremely important always to know the liquid level during cargo operations, most vessels have a back-up measuring system which can be used in the event of a breakdown in the primary system.

A high level alarm sounds when a tank is 95 percent full and again when the tank is 98 percent full. In the latter situation the alarm also exercises a control function by automatically closing the tank valve.

It is usual practice to retain a residue of cargo in a tank in order to keep the tank cool. A low-level alarm sounds when the liquid falls to a certain level to help prevent the tank being accidentally emptied.

(c) Pressure

Sensors monitor the vapour pressure and gauges indicate whether the pressure is too high or too low. If the pressure is too high the relief valves lower the pressure, but if the vapour pressure is too low the appropriate monitor shuts valves in the vapour line and stops the reliquefaction compressor.

A gas carrier has elaborate safety and control monitors located throughout the ship which are too numerous to mention. To illustrate the importance of such devices to operational safety two are mentioned.

Emergency loop

This is an emergency shut-down system. Cargo valves are held in the open position by air pressure which is delivered to each valve by a common pipeline which 'loops' around the ship. The air is released in an emergency, either by hand or automatically, thus closing all the cargo valves at the same time.

Gas detection sensors

These monitor spaces other than the cargo tanks, e.g. accommodation, duct keels, void spaces, control room, reliquefaction plant, and peak tanks. The sensors are located at various heights as cargo gases may be heavier or lighter than air. If gas is detected, alarms are sounded on the bridge and in the cargo control room.

Some operational procedures

The following guidelines are given for information purposes only and to provide an insight into the specialized nature of gas carriers. A wide variety of liquid gases are transported by sea and some of the gases may interact with each other, or with inert gas or air. The atmosphere within a tank must therefore be suitable for the cargo which is to be loaded.

Cooling a tank for loading

Before cooling takes place the tank must first be inerted and then purged with vapour compatible with the cargo to be loaded e.g. LPG vapour for an LPG cargo, methane for a LNG cargo, ammonia vapour for an ammonia cargo. Some vessels are equipped with deck storage tanks containing liquefied gas which can be used for the purging and cooling operations. On other vessels the purging vapour and cooling liquid must be taken from the shore.

The liquid used for cooling is introduced into a tank via the cooling spray rails. When the liquid droplets are sprayed into the tank vaporization of the droplets will occur. The energy, or heat, required for the vaporization process is obtained from the tank atmosphere and structure and thus the tank is cooled. The rate of cooling must be carefully controlled. The rate varies from ship to ship but on a fully refrigerated ship the rate will be within the limits of 4°C to 10°C per hour. The average time taken for the tank cooling process can be from 15 to 20 hours, depending on the tank size and the construction materials. As a tank cools, it will contract and thus it is necessary to 'bleed' nitrogen or inert gas into the void spaces around the tank to compensate for the increased volume of those spaces. The temperature of the tank structure must be carefully monitored to ensure that the temperature differential between various parts of the tank does not become excessive.

Pressure within a tank will increase due to the vaporization of the liquid droplets. This pressure is relieved by the reliquefaction process or by returning the vapour to the shore by a pipe which connects to a shore line at the manifolds and which is known as the vapour return line.

If a tank has been cooled at sea by using the reliquefaction plant and/or deck storage vessels, it will be necessary to spray liquid into the tank until a quantity lies in the bottom of the tank. Once a sufficient quantity of liquid is known to be established in the bottom of the tank loading can commence through the liquid loading line.

Loading an LPG carrier

The vessel would normally arrive at the loading berth with the tanks cooled to the required temperature, liquid loading lines cooled, reliquefaction plant running, and the cargo tanks at a slight positive pressure. Loading takes place through the liquid loading lines to the bottom of the tank to prevent a static electricity build-up. As the tanks fill, the vapour in the space above the cargo is reliquefied and led to the deck storage vessels. However, the vapour may become compressed and supersaturated and the rate of condensation on the tank structure may not be sufficient to prevent the tank pressure becoming too high. To prevent this occurring, cargo liquid may be led from the reliquefaction plant to the spray rails and reintroduced into the tank. The supersaturated vapour will then condense on the small droplets and the pressure will be relieved. In many cases the operation of the reliquefaction plant itself will prevent pressure building up and it will be unnecessary to reintroduce liquid into the tank.

The loading rate will, of course, vary from ship to ship but it can be in the region of 500 m^3 per hour for each tank being loaded. Each tank is loaded to 98 percent capacity and a high degree of operational efficiency is required for the topping-up process. Tanks are normally loaded to capacity as slack tanks can cause the following problems:
(a) cargo sloshing inducing tank stress
(b) excessive cargo pressure head on the longitudinal dividing bulkhead during rolling

The equalizing valves should be kept open for loading and discharging operations and closed for the sea passages. As with most types of gas carriers ballasting operations are concurrent with cargo operations.

Loading an LNG carrier

As previously mentioned, the low temperatures at which LNG is carried makes the installation of a reliquefaction plant economically and operationally impracticable, i.e. it is too expensive and difficult to reliquefy vapour which is already extremely cold. Consequently LNG ships employ the 'closed cycle' loading and discharging procedure as shown in Figure 7.2. The shore loading arms and vapour return line are connected to the liquid cargo line and vapour line at the ship's manifold.

Before loading starts the tanks must be dry and free of oxygen. The tanks will therefore have been dried with air which has been passed through a dehumidifier. The dewpoint of the air must be low enough to prevent condensation of water vapour on to the cold tank structure. The tanks will then have been inerted with nitrogen to prevent the formation of an explosive atmospheric mixture. Methane vapour is introduced from the shore until the inert gas is vented, usually to the atmosphere, and a suitable positive loading pressure is attained. A vessel may also arrive alongside gas-free and under an 80 percent vacuum. The vacuum is broken by the introduction of the shore methane vapour and the atmosphere inside the tank raised to a slight positive pressure. When methane is detected in the vented exit gas, cooling commences via the spray rails. Once a suitable level of liquid is formed in the bottom of the tank, loading is carried out as previously described.

During loading the cargo vapour must be returned to shore tanks via the ship's piping and the shore vapour return line. The rate of loading is

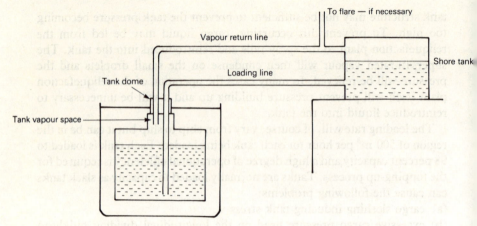

Figure 7.2   Closed cycle cargo system, LNG carrier

determined by the rate at which vapour is generated within the tank and the ability of the vapour return line to remove the vapour to shore.

Loaded passage

On LPG ships cargo vapour is withdrawn from the vapour space, passed through the reliquefaction plant, and the condensate returned to the tank. The removal of the vapour reduces the pressure in the tank and evaporation, or boiling, takes place from the liquid cargo. The evaporation process uses up latent heat and thus the cargo is cooled.

On LNG vessels the boil-off is either vented to atmosphere or used as propulsion fuel.

Discharging

Before commencing the discharge on refrigerated ships the liquid lines on deck must be cooled. The discharge may be aided by the use of ship's booster pumps which assist the individual tank pumps. Discharge is usually commenced from one tank only and once the system is observed to be functioning satisfactorily all the tanks are discharged simultaneously. Pressure must be maintained within the tank during all stages of the operation.

Ballast voyage

Some cargo is retained on board after discharge to be used as a coolant and to maintain pressure in the cargo tanks. An LNG carrier of 100,000 m³ cargo capacity would retain approximately 2000 m³ of cargo for cooling purposes.

224

Preparing a tank for entry after a cargo discharge

The following outline procedures should be adopted:

1 The puddle heaters should be used to boil-off any remaining cargo liquid. On LPG ships the vapour may be reliquefied and passed to a deck storage vessel. On some LNG ships the remaining liquid is boiled off by the introduction of heated vapour.
2 Isolate the tank from any reliquefaction plant and secure any cargo inlet or outlet valves in the closed position.
3 Carefully raise the tank temperature to the ambient (or surrounding) temperature.
4 Purge with inert gas to prevent any remaining cargo vapour forming a flammable mixture with air.
5 Ventilate with fresh air using fixed or portable fans of approved fire-proof construction.
6 Institute entry to enclosed spaces procedures.

Precautions must also be taken to prevent personnel coming into any sort of contact with toxic vapours which may be vented during the above operation.

Preparing a tank during a ballast voyage for a change of cargo

1 Boil-off any remaining cargo liquid.
2 Raise the tank temperature to ambient.
3 Purge from LPG or LNG to inert gas, or from ammonia to dry air.
4 Purge the inert gas with vapour compatible with the next cargo (the vapour is obtained from the deck storage vessels).
5 Cool the tank to the required loading temperature.
6 Ensure that the void spaces surrounding the tank are inerted.

Safety procedures

Safety procedures are of such paramount importance on gas carriers that it would be inappropriate for the hazards to be discussed in a book of this nature. It may be sufficient to say that a gas vapour can be flammable, explosive, toxic, asphyxiating, corrosive, poisonous, and can inflict frostbite or other injuries on personnel. Specialized information should be consulted before evaluating or discussing safety procedures on gas carriers.

225

C   Chemical Cargoes

IMO Codes

The 'Code for the Construction and Equipment of Ships Carrying Dangerous Chemicals in Bulk' applies to ships built before July 1986; the International Code for the Construction and Equipment of Ships Carrying
Dangerous Chemicals in Bulk' (the International Bulk Chemical Code – IBC
Code) applies to ships built after June 1986.

The purpose of the codes is 'to recommend suitable design criteria,
construction standards and other safety measures for ships transporting
dangerous chemical substances in bulk'.

The codes apply to bulk cargoes having:

(a) significant fire hazards

(b) other significant hazards

A list of such liquids is contained within the codes which, at present are limited
to tank ships. The hazards which are considered are fire, health, water pollution, air pollution, and reactivity.

It is possible that damage to a vessel carrying dangerous chemicals in bulk
may lead to an uncontrolled release of cargo. The siting of the cargo tanks in
relation to the ship's side and bottom is therefore dependent on the potential
hazardous effect of the cargo on the environment. Three degrees of physical
protection are employed:

Type I is the highest standard of protection and is required for those
substances possessing the greatest environmental hazard.

Types II and III have reduced standards for substances of progressively lesser
hazard.

The protection is mainly afforded by siting the cargo tanks in specific
locations within the ship in order to minimize the effect on cargo containment
from external damage to the ship.

The codes also contain stringent requirements for cargo transfer and vent
systems, control equipment, safety equipment, and special requirements for
certain cargoes.

British Regulations

M1237 draws attention to the Merchant Shipping (Chemical Tankers) Regulations 1986, operative 1 July 1986, which make the IBC Code mandatory for
UK ships throughout the world and for foreign ships in British waters. M1169
draws attention to the survey requirements for UK chemical carriers and notes
that, in addition to an initial survey for the issue of a Certificate of Fitness and a
periodical survey for the renewal of the Certificate, annual surveys and at least

one intermediate survey are also required during the validity of the Certificate (the survey periods are the same as those for gas carriers as described earlier in the book). The Fitness Certificate lasts for five years and M1169 should be studied for the details of the surveys.

A 'Certificate of Fitness' gives the following information:

(a) Name of ship
(b) Port of registry
(c) Ship type
(d) Cargoes which the ship is permitted to carry
(e) The conditions of carriage
(f) Any authorized exemptions permitted by a code

Tanker Safety Guide (Chemicals)

The guide is published by the ICS and its purpose is to provide those serving on ships carrying bulk liquid chemicals with recommendations on recognized good practice in operational matters. The guide consists of four volumes in a loose-leaf format. Volume 1 contains general information pertaining to overall precautions, fire hazards and precautions, cargo and ballast operations, tank cleaning and gas freeing, health and chemical hazards, static electricity, and fire-fighting and life-saving procedures. The other volumes contain cargo data sheets.

Information is obtained by first referring to the 'Index of Chemical Names' which is located at the back of Volume 1. The index gives the number of the data sheet which relates to the cargo in question and from the sheet the hazards associated with the particular cargo can be readily identified. Many chemicals have more than one name in common use and thus the chemical names of the numerous products are used in preference to trade names or synonyms. The information on the data sheets is listed under the following headings:

appearance
odour
the main hazards
emergency procedures in the event of accidents
fire and explosion data
chemical data
reactivity information
health data
physical properties
handling and storage recommendations

The ICS also publishes a booklet, *Safety in Chemical Tankers*, which highlights some of the particular hazards of sailing on such vessels.

227

Chemical Hazard Response Information System (CHRIS)

This system, which is operated by the United States Coastguard, is designed to provide information essential for proper decision-making during emergencies involving the water transport of hazardous chemicals. CHRIS operates in the USA and consists of regional contingency plans, a hazard-assessment computer system, Coastguard organization units, and four reference manuals. The manuals should be studied by all personnel involved in the water transport of chemicals as, in addition to emergency operational procedures, the manuals provide basic information which can be used to improve safety standards.

Medical guides

All chemical tankers should carry the following:

Medical First Aid Guide—for Use in Accidents Involving Dangerous Goods (MFAG), published by IMO
'Chemical Supplement' to *The Ship Captain's Medical Guide* (SCMG), published by the Department of Transport

M1050 describes the relationship between the two volumes.

Bulk chemical carriers

These ships are often known as 'parcel carriers' and are designed to carry small 'lots' of liquid chemicals or parcels ranging from a few hundred to several thousand tonnes each. Deep-sea parcel carriers are generally between 10,000 and 35,000 tonnes deadweight. In appearance the ships look like product tankers but the cargo compartments ensure a very high degree of cargo segregation and the cargo-handling arrangements are more elaborate. The number of tanks can vary from approximately 20 to 40 and it is not unusual for a ship to carry 30 separate parcels. Tanks may be constructed of stainless steel or may be coated with epoxy, zinc or other chemical-resistant paint.

On modern vessels each tank is fitted with its own stainless steel hydraulically driven submerged cargo pump with a discharge rate varying from approximately 50 to 1000 m³ per hour. The individual discharge line from each tank connects with one of the deck main pipelines, of which there may be six. The individual branch line system permits great flexibility in cargo handling and reduces the amount of line cleaning required between grades. In general, the cargo-handling operations are similar to that of product tankers.

The characteristics of the chemicals carried vary considerably. About 50 percent of the liquids carried are petrochemical such as benzene, glycols, toluene and xylene. The remaining cargoes can be subdivided into three groups:

(a) natural products, vegetable oils such as soya bean and linseed, and animal fats such as lard and fish oils
(b) inorganic chemicals, such as sulphuric acid, phosphoric acid, caustic soda
(c) specialized liquids, such as lubricating oils

Chemical cargoes must be transported with a high degree of operational efficiency to ensure that no contamination of very valuable cargo occurs.

Tank materials and coatings

Stainless steel

Some cargoes are highly corrosive to the mild steel with which most ships are constructed and stainless steel tanks are used to contain such cargoes. Many grades of stainless steel are available and a grade which is compatible with many different chemicals is usually chosen for a particular tank. Stainless steel's resistance to corrosion depends upon a thin passive surface layer of chromium oxide. This layer can be damaged by halides (compounds of halogens) such as chlorides and fluorides and by certain acids, e.g. hydrochloric. Sea water ballast must never be put into stainless steel tanks as sea water contains chloride. For the same reason such tanks should only be washed with fresh water.

Coatings

Mild steel is resistant to most chemicals. Unfortunately it rusts easily and most chemicals are contaminated by rust. Mild steel tanks are therefore coated for the following reasons:

1 Corrosion prevention to avoid cargo contamination.
2 The coated surface aids tank cleaning.
3 Gas freeing is made easier as no corrosion sludge traps cargo gas.
4 Stains are easier to detect.
5 Safety is improved because of compatible coatings with cargo.
6 The risk of contaminating subsequent cargoes is decreased.

However, as the completeness of a tank coating can never be guaranteed, coatings should never be relied upon to protect the tank. Each parcel should be compatible with the tank coating, the tanks should be frequently inspected for failures in coatings, and any damage should be repaired immediately. No one coating is suitable for all parcels and a ship may have all or some of the following coatings applied to the cargo tanks.

Epoxy   Resistant to alkalis, amines (derived from ammonia), weak acids, glycols, vegetable and animal oils, mineral oils, and sea water ballast. It is unsuitable for alcohols such as ethanol and methanol, esters and ketones. Aromatics, such as benzene and toluene, are sometimes carried in epoxy-coated tanks but the paint has only limited resistance to aromatics. Epoxy coating can be softened by some chemicals and thus can absorb small amounts of the cargo. If this occurs the tank must be ventilated and washed before the next cargo is loaded.

Zinc silicate   This coating is highly resistant to strong solvents such as ketones and esters and to aromatics such as benzene and toluene. It will not withstand acids or alkalis, including sea water. The pH value (an indication of the acidity of a solution) must be within the range 5.5 – 11. Animal and vegetable oils with a free fatty acid (FFA) content of more than 2.5 percent should not be carried in zinc silicate coated tanks. If halogenated compounds such as carbon tetrachloride are to be loaded, zinc-protected tanks must be absolutely dry before loading commences. Many of the compounds hydrolize in the presence of moisture and the acids which form will damage the tank coating.

Polyurethane   The increased performance of this type of coating has made it suitable for chemical carriers since the late 1970s. Those chemicals which are transported in epoxy-coated tanks are also compatible with polyurethane. The high gloss finish of polyurethane sheds cargo easily and gives it an advantage over the matt finish of epoxy.

Phenolic   This is modified epoxy which gives better resistance to solvents and aromatics. Phenolic coatings give excellent resistance to most chemicals but the paint is more expensive than conventional epoxy.

Tank washing

The tank-washing process on chemical tankers is closely similar to the water-washing operation on oil tankers. The washing medium is usually fresh or sea water, depending on the compatibility of tank coating, previous cargo, and expected cargo. The washing is carried out by means of fixed or portable tank-cleaning machines and the residues are stripped to a slop tank. The disposal of slops is governed by Annex II (Regulations for the Control of Pollution by Noxious Liquid Substances in Bulk) of MARPOL 73.

Substances are divided into four categories A to D depending upon their environmental effect if discharged into the sea during tank-cleaning or deballasting operations.

The Annex gives the requirements for the discharge of residues only into shore reception facilities unless certain conditions are complied with. However, under no circumstances is a discharge of residues containing noxious substances permitted within 12 miles of the nearest land and in a depth of water of less than 25 metres. Stricter regulations apply in 'special areas' which are the Baltic and Black Seas.

Cargo Record Book

Regulation 9 of Annex II to MARPOL 73 states that the Cargo Record Book must be completed, on a tank-to-tank basis, whenever any of the following operations are carried out:

1 Loading.
2 Discharging.
3 Cargo transfer.
4 Transfer of residues to a slop tank.
5 Tank cleaning.
6 Transfer from slop tank.
7 Ballasting of cargo tanks.
8 Transfer of dirty ballast water.
9 Any permitted discharge into the sea.

Any discharges of cargo into the sea for the purposes of securing the safety of a ship or saving life at sea, or discharges resulting from damage to a ship or its equipment, or the discharge of anti-pollution mixtures containing noxious substances, must also be recorded in the book.

Each operation entry should be signed by the officer in charge of the operation and the Master is required to sign each page. The book is liable to be inspected at any time while the vessel is in port.

Equipment items

The construction code and operational manuals should be consulted to obtain the relevant requirements for specific items of equipment. A few items of general interest are noted below.

Tank high level alarm

Each cargo tank is fitted with an alarm which indicates when there is imminent danger of the tank being overfilled.

Tank overflow control

An approved system must be fitted to cargo tanks to ensure that cargo cannot overflow on to the deck or overboard during the loading operation.

Tank vent systems

All cargo tanks should be fitted with a system which is suitable for the cargo being carried. Vent systems must be of a design which minimizes the possibility of cargo vapour accumulating about the decks, entering accommodation and machinery spaces, and other spaces containing ignition sources.

Cargo valves

These are usually operated hydraulically or pneumatically by remote control. However, when possible a local inspection should be carried out to ensure that valves are in the required open/closed position.

Remote readings

There are three basic remote readings required for cargo operations:
(a) tank level
(b) cargo temperature (of heated cargoes)
(c) cargo line pressure

Other remote readings may include open/closed position of valves, pump pressures, pump speeds, and the inert gas system. Most chemical tankers have a cargo control room where alarms, remote gauges, and read-outs are centralized and from which cargo operations are directed.

Toxic vapour and gas detection

As one would expect, stringent provisions are made for the detection of noxious vapours on chemical carriers. The IMO code requires the provision of instruments for testing the possibilities of both toxic and flammable gas concentrations. It is normal practice to have a fixed gas detection system which is similar to LPG/LNG carriers and portable detectors which are used before entering enclosed spaces. Enclosed space entry procedures on chemical tankers are of a very high standard. If a large escape of toxic gas does occur the ship's accommodation area can be made completely gas-tight.

Inert gas plant

The 78 Protocol to SOLAS 74 requires the installation of an inert gas system on chemical tankers of 20,000 tonnes deadweight and above. 'Intermin Regulations for Inert Gas Systems on Chemical Tankers Carrying Petroleum Products' can be found in the IMO publication on *Inert Gas Systems, 1982.*

Cargo information required before loading

1 The correct chemical name of the cargo should be provided so that the appropriate data sheet in the *Tanker Safety Guide (Chemicals)* can be consulted.
2 Quantity in weight.
3 Required quality control. Contamination is measured in parts per million (ppm); thus tanks and pipelines must be practically spotless.
4 Specific gravity. This is required in order that an estimation can be made of the probable volume that the weight quantity will occupy.
5 Temperature. This is required for two purposes:
   (a) The loading temperature is used in conjunction with the specific gravity to obtain the probable volume of the particular parcel.
   (b) The temperature at which the cargo is to be carried will indicate if heating will be required on passage. Some chemicals will solidify or polymerize if a certain temperature is not maintained. Polymerization is a chemical reaction in which small molecules combine into larger or very large molecules which contain thousands of the original molecules. Thus a free-flowing liquid can become a viscous liquid or even a solid.
6 Compatibility. Certain chemicals react with other chemicals and thus may not be stowed in adjacent compartments.
7 Tank coating compatibility. The tank coating must be suitable for the proposed cargo.
8 Corrosive properties. This will also indicate the required tank coating and also possible damage to ship fittings.
9 Electrostatic generation. Some chemicals can accumulate static. The principles which apply to hydrocarbon cargoes should be applied to chemical static accumulators.
10 Fire and explosion data. It has been previously noted that 50 percent of the chemicals which are transported are derived from hydrocarbon oils and thus the fire hazards are similar to those which pertain to petroleum products.
11 Toxicity. Chemicals which emit highly toxic vapours require closed ventilation and ullaging systems.

12 The health hazards of the particular parcel.
13 Reactivity.
14 Action to be taken in the event of particular emergencies.

Most of the above information, and additional essential information, can be found on the chemical data sheets in the safety guide.

Reactive cargoes

Special attention must be paid to the possibility of certain cargoes reacting with their environment. There are three basic types of reaction:
(a) Reaction with air, or self-reaction when only the chemical itself is involved
(b) Reaction as a result of mixing with, or contamination by, another chemical
(c) Reaction due to contact with water

Reaction can be avoided by adopting some, or all, of the following procedures:

1 Add an 'inhibitor' to the cargo. This is the term used to describe a compound which causes a chemical to become stable by slowing down or stopping a chemical change. Thus, an inhibitor can be added to a chemical to stop polymerization occurring.
2 Handle cargo in a closed system to prevent mixing with air.
3 Blanket the cargo with an inert gas such as nitrogen, i.e. purge the tank with inert gas of very high purity and keep the tank inerted.
4 Use an approved heating system.
5 Ensure that the materials in the cargo-handling system are compatible with the chemical.
6 Ensure that the tank is corrosion free and that the coating is compatible.
7 Do not stow reactive cargoes adjacent to each other.
8 Stow water reactive chemicals in a double separation compartment.
9 Avoid contamination
10 Use a suitable vent system.

It is important to remember that reaction can cause a fire or explosion, release dangerous vapours, produce heat, increase tank pressure, increase the health hazards, change the nature of the cargo, and lower the cargo quality.

Precautions to be observed during cargo operations or tank washing in port

1 A procedural plan should be adopted which must be agreed to by ship and shore personnel.

2 Complete the ship/shore check list.

3 Check that the emergency decontamination unit is functioning properly, e.g. water in the showers, eyewash and antibiotics available.

4 All safety equipment such as protection suits and compressed air breathing apparatus available and functioning.

5 Fire precautions observed, e.g. fire hoses rigged on the offshore side and pressurized. Ensure that a suitable system is available for use with a particular cargo.

6 Moorings kept tight and constantly attended.

7 Emergency towing-off wires constantly adjusted so that tugs can easily hook on to the wires.

8 Suitable access to be provided. When possible access should be via the poop.

9 Adequate illumination should be provided.

10 No unauthorized persons allowed on board and all unnecessary ship's personnel kept out of the cargo working area.

11 Notices should be displayed at the ship's access:

(a)

Warning

No Naked Lights

No Smoking

No Unauthorized Persons

(b) When engaged in handling cargoes which present a health risk:

Warning

Hazardous Chemicals

12 Warning notices should also be posted inside the accommodation instructing personnel to stay within the accommodation.

13 Tank deck scuppers plugged.

14 All pipe joints and flanges checked for tightness.

15 No unauthorized craft alongside.

16 Suspend operations during inclement weather such as thunder storms or in conditions which do not permit cargo vapour to blow away.

17 If cargo vapours are toxic or a health hazard:

(a) prevent entry of vapour to the accommodation

(b) air conditioning on internal cycle operation

(c) breathing apparatus worn when personnel are outside the accommodation

(d) gas detection system to be operating correctly

(e) use a closed ullaging system

(f) use a closed vent system

18 Red flag or red light exhibited when handling petrochemicals.

19 Ensure that boiler tubes are not blown and that funnel spark arrestors are in good condition.
20 Galley precautions observed.
21 All relevant information given to the personnel who will be involved in the operation.
22 The Emergency Squad available to deal with incidents such as spillage at the cargo manifold.

Fire protection

Chemical carriers require sophisticated fire detection and fighting equipment and techniques due to the hazardous nature of the cargoes which are transported. The basic requirements are found in SOLAS 74 and in the IMO construction and equipment code. The IMO code requires chemical carriers to be equipped with suitable fire-extinguishing equipment for all products which are to be carried and if products evolve flammable vapours the equipment should include an approved fixed extinguishing system. Such systems may be:

A   alcohol foam deck system
B   regular foam deck system
C   water spray
D   dry chemical

The increased sophistication of chemical carriers now enables such ships to carry hundreds of different chemicals and it has become increasingly difficult to design fire-fighting systems capable of dealing with all cargo fires. The most common system is 'two-fold', e.g. an external system to deal with spillage fires on deck, associated with an internal preventative system inside the cargo tanks.

It is essential that the data sheets in the *Tanker Safety Guide (Chemicals)* be consulted to ascertain the correct extinguishing medium for particular cargoes. Chapter 12 in Volume 1 of the guide deals with fire fighting and the chapter should be studied carefully. Particular attention should be paid to the reactions which chemicals may have with fire-extinguishing mediums, e.g. some chemicals destroy water-based foam; some chemicals react with water and produce heat.

Personnel protection

A chemical cargo may:

1 Be corrosive and destroy any human tissue which comes into contact with it.

2 Be poisonous and may enter the body by several methods.
3 Be toxic and inhalation may cause brain damage, damage to the nervous system and vital organs.
4 Give off flammable gas.

Many chemicals can cause death and thus personnel protection on chemical tankers is of a very high standard.

The IMO construction and equipment code requires the personnel involved in cargo operations be provided with suitable protective clothing and equipment which covers 'all skin so that no part of the body is unprotected'. Protective equipment includes:

(a) protective suit (in the form of a boiler suit) made in resistant material with tight-fitting cuffs and ankles
(b) helmet
(c) boots
(d) gloves
(e) face shield or goggles
(f) large apron

When handling products which present inhalation problems the above equipment is supplemented by a suitable breathing apparatus. The code requires that 'ships carrying toxic cargoes should have on board sufficient but not less than 3 complete sets of safety equipment' in addition to the requirements of SOLAS 74. Each set should consist of:

(a) one self-contained air-breathing apparatus
(b) protective clothing, boots, gloves and tight-fitting goggles
(c) steel cored rescue line with harness
(d) explosion-proof lamp

An air compressor and spare cylinders must also be carried and all compressed air equipment should be inspected at least once a month and tested yearly by an expert.

When toxic chemicals are being carried respiratory equipment should be available for all personnel on board. The duration of such equipment must permit escape from the ship in the event of a major accident.

The ships are also provided with suitable medical first aid equipment which includes oxygen resuscitation apparatus and antidotes for the cargo carried. Decontamination centres which contain showers and eyewash facilities are situated on deck.

All the above equipment must be suitably located about the deck.

Specialized information should be consulted to obtain further information on the health and safety hazards pertaining to chemical cargoes.

Further reading

CORKHILL, M. *Chemical Tankers* (Fairplay: London, 1981).
FFOOKS, R. *Natural Gas by Sea* (Gentry: London, 1979).
WOOLCOOT, T. W. V. *Liquified Petroleum Gas Tanker Practice* (Brown: Glasgow, 1977).
WOOLER, R. G. *Marine Transportation of LNG* (Cornell Maritime; Cambridge, Maryland, 1975).

Handbook

Gas and Chemical Ships Safety, Bureau Veritas (Lloyds of London Press, 1986).

ICS and OCIMF guide

Ship to Ship Transfer Guide (Liquefied Gases).

IMO Guide

Index of Dangerous Chemicals Carried in Bulk.

OCIMF guide

Safety Inspection Guidelines and Terminal Safety Check List for Gas Carriers.

Journal

MURRAY, J. W. 'The practical operation of a modern LNG/LPG carrier, *Seaways,* October 1982, pp. 2–6.

8

Cargoes Which Have a Large Potential for Shifting

A   Deck Cargoes
B   Grain Cargoes
C   Solid Bulk Cargoes

A Deck Cargoes

Merchant Shipping (Load Line) (Deck Cargo) Regulations 1968. SI 1968 No. 1089

These regulations were made under the MS (Load Lines) Act of 1967 and came into operation on 29 July 1968. They state the requirements which must be complied with when cargo is carried in uncovered space on the deck of a ship, i.e. deck cargo. The regulations are in two parts.

Part I. Requirements which apply to ALL deck cargoes

The distribution and stowage of deck cargo must be carried out with due attention being given to the factors mentioned below:

1 Avoid excessive loading and have regard to the strength of the deck and the supporting structure of the ship. Decks have been set down due to excessive local and overall weights, and hatch covers have been damaged due to the weight of cargo loaded on them.
2 The ship must retain adequate stability at all stages of the voyage. Particular attention should be paid to:
  (a) the vertical distribution of the deck cargo;
  (b) expected wind moments which may be produced by strong winds encountered within the trading area (the windage area, its centre of gravity, and the lever to mid-draught can be found in the *Stability Information Booklet*);

    (c) losses of weight within the ship, e.g. the rise in the position of the ship's centre of gravity due to the consumption of fuel in the double-bottoms;

    (d) possible gain in the weight of deck cargo which would also cause a rise in the position of the ship's centre of gravity, e.g. caused by absorption of water into the cargo or by excessive icing on the deck, superstructure, and cargo.

3 The weather-tight and water-tight integrity of the ship must not be impaired. Special attention should be paid to the protection of ventilator and air pipes (M 1051 refers to an incident in which deck cargo sheared the air pipe to a deep tank).

4 The height above the deck should not interfere with the navigation or working of the ship, e.g. containers should not be stowed so high as to impair the keeping of an effective lookout.

5 Access to the ship's steering gear arrangements, including the emergency steering arrangements, must not be obstructed.

6 The cargo must not obstruct or interfere with crew access to accommodation and working spaces or obstruct any opening and prevent it being easily secured weather-tight.

Deck cargo must be secured so as to ensure that there will be no movement of cargo in the worst weather that can be expected on the voyage. The lashings and all fittings used for the attaching of lashings must be of adequate strength to be able to withstand the rigours inflicted upon the ship and cargo by such weather conditions.

Some ships which regularly carry deck cargoes have a passage constructed on or below the deck which carries the deck cargo, to provide access for the crew between their quarters and the working areas of the ship. On ships without such a passage, a walkway must be fitted over the deck cargo and effectively secured to provide safe and efficient access for the crew. The walkway must not be less than 1 m wide and it must have a set of guard rails or wires on each side which are supported by stanchions securely fitted to the walkway at intervals not exceeding 1.5 m. The guard rails or wires should be to a height of not less than 1 m and each set should consist of three courses. No opening below the lowest course should exceed 230 mm in height and no opening above that course should exceed 380 mm in height.

Part II. ADDITIONAL requirements applicable to timber deck cargoes

Part II is divided into Sections A and B. Section A applies to ships which are not marked with timber load lines or to ships that have timber load line marks but which are loaded within the limits of ordinary load lines.

When the ship is in a winter period the deck cargo must be stowed so that at no point throughout its length the height of the cargo above the weather deck exceeds one-third of the extreme breadth of the ship. Note that the height is not taken as an 'average' or a 'mean'. The regulation emphasizes that the maximum height at any point must not exceed one-third of the beam.

A walkway must always be provided on top of the timber even if the ship has a permanent passageway of the type described in Part I. The walkway must be constructed to the specification set out in Part I.

The cargo must be compactly stowed and secured throughout its length by a system of overall lashings of adequate strength. Efficient arrangements, which must be readily accessible at all times, must be provided for the release of lashings and fittings.

Uprights, which are sufficiently strong for the purpose, must be fitted if the nature of the timber is such that uprights are necessary to ensure a compact and secure stow. The uprights must be secured in position by angles or metal sockets. The spacing of the uprights must take into account the nature and length of the timber so that efficient support is provided. However, the space between any two uprights must not exceed 3 m.

Section B applies to ships which are marked with timber load lines and which are loaded accordingly. The requirements of Section B are in addition to those of Section A.

A careful note should be made of the regulation which refers to a minimum height of cargo, i.e. when timber deck cargo is stowed in any well it must be 'stowed as solidly as possible so as to extend over the entire available length of the well to a height not less than the standard height of a superstructure other than a raised quarter deck'. Thus in addition to the maximum height in summer of a safe height (Part I) and the maximum height in winter of one-third of the beam Part IIA), the cargo must be stowed to a minimum safe height in both seasons (Part IIB). If a ship has no superstructure aft the timber must be stowed so as to extend over the entire available length between the superstructure and the after end of the aftermost hatchway.

The timber must be efficiently secured throughout its length by independent overall lashings spaced not more than 3 m apart (thus if one independant lashing breaks it will not affect the other independent lashings). The lashings must be secured to eye plates which are attached to the sheer strake or to the deck stringer at intervals of not more than 3 m. The distance from an end bulkhead to the first eye plate must not be more than 2 m. Where there is no bulkhead the eye plates and lashings must be located at distances of 0.6 and 1.5 m from the ends of the timber deck cargo. Students may find it useful to make a sketch of the lashings arrangements, as in Figure 8.1.

The lashings must be made of close link chain of a size not less than 19 mm or of flexible wire rope of an equivalent strength. The lashings must be fitted with sliphooks and turnbuckles in positions which are accessible at all times and wire rope lashings must be fitted with a length of long link chain to enable the length of the lashings to be regulated.

Vessel with superstructure forward and aft

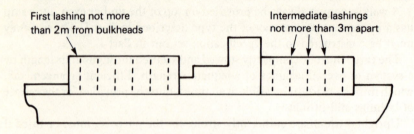

First lashing not more
than 2m from bulkheads

Intermediate lashings
not more than 3m apart

Vessel with no superstructure aft

Cargo must be stowed to aft end of aftermost hatch

First lashing
0.6m from end

Second lashing
1.5m from end

Other lashings as above

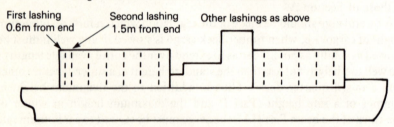

Figure 8.1   Lashings on timber deck cargoes, vessel loading to timber load line

When timber is in lengths less than 3.6 m the spacings of the lashings must be suitably reduced.

A student who closely studies the above regulations will observe that the regulations contained in Part I are written in 'broad' or general terms. Deck cargoes can be of such a diverse nature that it would be difficult to lay down 'hard and fast' rules which apply to all deck cargoes. However, when dealing with timber deck cargoes it is possible to be more specific and thus Part II contains detailed securing arrangements. Students must remember that Part I also applies to timber deck cargoes and the general provisions must be complied with.

A timber deck cargo stowed solidly in wells has an effect which is similar to that of raising the height of the freeboard deck. Thus the assigned freeboard may be reduced and the ship loaded to a deeper draught. A solidly stowed, properly secured timber deck cargo means that:

1 Water cannot flow freely on the weather deck.
2 Greater protection is given to closing appliances, such as hatch covers, which protect the weather-tight and water-tight integrity of the ship.
3 The timber compensates for the loss of reserve buoyancy caused by the reduced assigned freeboard. It is, therefore, important that the timber cargo is solidly stowed to the minimum height specified in the regulations.

M687 should be closely studied as it gives further guidance for the transporting of timber deck cargoes. The notice gives the definition of timber as:
'(a) loose timber, i.e. individual planks of sawn timber, logs or pit-props; or
(b) packaged timber, i.e. bundles of sawn timber, consisting of either
    (i) similar sized planks forming a regular cuboid—(regular packaged timber) or
    (ii) random sized planks which do not form a regular shape—(irregular packaged timber).'
The notice refers to the fact that the transverse shift of stowed timber is seldom the sole cause of a dangerous list. An inadequate standard of stability, imprudent stowage, and unsecured deck openings will aggravate the effects of a list caused by a shift of deck cargo. The notice also refers to the general rule that not more than one-third of the weight of timber carried should be stowed on the open deck.

M687 refers to the IMO publication, *The Code of Safe Practice for Ships Carrying Timber Deck Cargoes*, and M1279 draws attention to the Supplement to that code (the Supplement follows Annex D in the reprint of the 1981 edition which was published in 1987). The Department of Transport recommends that the requirements and advice contained in the publications should be observed.

Code of Safe Practice for Ships Carrying Timber Deck Cargoes

The IMO code should be studied in conjunction with the Merchant Shipping Regulations as similar provisions are contained in both publications. However, some of the additional IMO requirements are noted below.

Uprights should extend above the outboard top edge of the cargo, they should be fitted with a locking pin, and each port and starboard pair of uprights should be linked by athwartships lashings.

If a device which is capable of quick cargo release is fitted, the design of it should ensure that it cannot be accidentally released or activated.

Lashings should be capable of withstanding an ultimate load of not less than 133 kN (13,600 kP). The minimum ultimate load of the ancillary components is higher than that of the lashings. M929 refers to the fact that the ultimate strength of 19 mm close link chain is 13,600 kP and thus the chain may continue to be used. A flexible wire rope of equivalent strength is one which has a SWL of not less than 2.7 tons.

The spacing of lashings should be determined by the maximum height of the timber in the vicinity of the lashing. The following spacing is stipulated:
'(a) for the height of 4 metres (13 feet) and below the spacing should be 3 metres (9.8 feet);
(b) for the height of 6 metres (19.6 feet) and above the spacing should be 1.5 metres (4.9 feet);
(c) at intermediate heights the average spacing should be obtained by linear interpolation.'

In addition to the walkway specified in the Merchant Shipping Regulations, a lifeline of wire rope with a stretching device should be set up taut as near as practical to the centreline of the ship.

The IMO publications should be read in full as they contain much practical advice with regard to the stowing and securing of timber deck cargoes. Students studying for examinations should note that the words 'stowing' and 'securing' refer to different aspects of the transporting operation.

Annex C of the code gives suggested practices which pertain to particular timber cargoes, e.g.

1 The stowage of packaged timber.
2 The securing of heavy logs additional to that given in the code.
3 The stowage of pulp wood and pit props.

Annex D refers to 'Stowage and securing of deck cargoes of cants'. Cants are logs which have been slab cut, i.e. they are cut lengthwise from a tree trunk into thick slabs which have two opposing, parallel flat sides, a third side which is usually sawn flat, and a fourth side which is rounded. Cants are usually made up into rectangular 'bundles' which are secured by steel bands. The average length of the bundles is about 4–5 m with the maximum length being approximately 8 m. The nature of the bundles makes a solid, compact stow difficult to achieve and the guidelines contained in Annex D should be considered. The sketches in the Annex should be studied by mariners unacquainted with the transporting of timber as they clearly illustrate some of the securing arrangements.

General guidelines for loading timber deck cargoes

1 All securing equipment should be laid out on the main deck some time before loading commences. The IMO code recommends that a visual examination of lashings be carried out at intervals not exceeding 12

months. However, the prudent officer should inspect all the equipment before it is used. All the equipment, such as shackles, should be appropriately marked and the necessary certification should be in order. It may be necessary to weld additional securing points, such as eye plates, to the deck at appropriate locations. The outboard ends of lashings should be secured to eye plates before loading commences and temporarily secured in a location where they will not be overstowed. It may be necessary to hang the lashings over the side of the ship on temporary rope pennants so as to keep the deck clear for loading.

2 Provision should be made to protect ventilators, air pipes, sounding pipes, winches, and similar deck appliances. The cargo should be stowed so that soundings of compartments can be carried out during the voyage. Ensure that freeing ports are operating in a satisfactory manner.

3 Provision should be made to enable fire hoses to be rigged should that become necessary on passage and fixed fire-fighting appliances, such as the fire main, should be protected.

4 When necessary, e.g. for packaged timber, dunnage of rough timber should be laid on the deck. Lengths of 3 × 12 cm dunnage can be laid at 1 m spacing. If the underdeck framing runs athwartships the dunnage should be laid diagonally across several frames so as to spread the load of the timber.

5 M687 emphasizes that all hatch covers, weather-tight doors, and means of securing air pipes and ventilators that are situated within the deck cargo area must be made secure before loading commences.

6 Inspect packaged timber on the quay and reject any packages that are bound too loosely.

7 Neatly stow the timber so that, if possible, the timber interlocks within each layer. Do not 'stack' timber piece on piece as this will not ensure a compact stow.

8 Chock each tier before commencing another layer or tier. Any gaps at hatch coamings or around deck appliances should be filled with loose timber or chocked off. Keep a firm loading surface throughout the stow and, if necessary, dunnage between layers or tiers.

9 Any openings around deck appliances or mast houses should be fenced off to prevent personnel falling into gaps in the cargo.

10 Properly constructed and secure ladders or steps should be provided where necessary from the top of the cargo to the deck, especially in the vicinity of walkways.

11 In general, stow fore and aft to facilitate the securing arrangements but stow athwartships when it is necessary to 'fill in' around mast houses.

12 Distribute the weight throughout the stow to ensure that heavy timber is not stowed above or adjacent to light timber.

13 M687 recommends that the height of the timber should not be excessive

as the risk of shifting increases with the height of the cargo. The heeling effect of a shift is also greater when the cargo is stowed to an imprudent height.

14 M1110 reminds Masters 'of the importance of using slip hooks or other cargo-securing appliances of a satisfactory type, and maintaining them in good condition'. Slip hooks should be examined to ensure that they are incapable of working loose during the voyage. Straight tongue hooks of the type shown in Figure 8.2 should not be used.

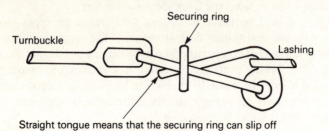

Figure 8.2   Non-approved slip hook

15 The outboard edges of the stow should contain no protruding cargo which would interfere with the vertical lead of the lashings. The top layer or tier should not overhang the vertical face of the cargo and it may even be advisable to 'step in' the top tier so as to provide a good lead for the lashings.

16 Annex C recommends that rounded angle pieces of suitable material should be placed along the top outboard edge of the stow. The angles will spread the load stress of the lashings, reduce chafing, and permit movement of the lashings during the tightening and securing of the cargo.

17 A practical test to check the vessel's stability before loading is completed has been used to good effect on some timber carriers. When one-third of the deck cargo remains to be loaded slings of timber are lifted from the jetty simultaneously by several of the ship's derricks and swung inboard. The resulting movement of the ship may indicate any undue stiffness or tenderness. Thus, any undesirable stability characteristics can be corrected by checking the stability calculations and carrying out corrective procedures before the ship sails.

18 The IMO supplement recommends that after the initial securing of the cargo, all tightening devices should be capable of further tightening for use on the voyage. The cargo will probably settle at the beginning of the voyage and the lashings will therefore require tightening.

19 The lashings and other securing arrangements should be frequently inspected on passage and all the lashings should be kept taut at all times.

20 A record of inspections and tightening of lashings should be kept.
21 On a long voyage it may be necessary to re-coat wire lashings with a suitable anti-corrosion compound.

Some methods to facilitate the jettisoning of timber deck cargoes

Jettisoning cargo is only carried out when a vessel is in extreme peril as it is an extremely hazardous operation. Cargo should only be jettisoned as a last resort to prevent a vessel sinking or to prevent loss of water-tight integrity. The emphasis of the regulations is to keep the stow intact, even during the worst possible weather conditions that can be anticipated. However, some procedures have been recommended which could make the operation slightly less hazardous.

1 Slip hooks and other fittings may become jammed during a voyage and it may be difficult to release them in an emergency. This problem can be reduced by fitting additional slip hooks back to back with the primary slip hooks in the overall lashings (M1110).
2 A set of wire or metal cutters should be provided so that strained or jammed lashings can be cut away (M1110).
3 It is possible that the jettisoning operation would be accomplished with more safety if a system of additional temporary lashings which could be released from a safe position is used (M1110). Such a system could be the use of a 'wiggle wire' which is described on page 12 of the supplement to the IMO code. The wiggle wire holds a stow together in the manner of a shoe lace. One end of a wire is secured forward and the rest of the wire is led aft in a continuous zig-zag over the top of the stow via a series of snatch-blocks. The wire can be heaved tight by the use of a cargo winch or aft mooring winch and released at the appropriate moment from a safe position aft.
4 Some quick release devices can be fitted to the securing arrangements but the IMO code stipulates that such devices must not be capable of being accidentally released or activated. There are no strict guidelines as to the advisability of using quick release devices; the IMO codes stipulates, 'if fitted'.
5 Releasing devices should be located along the centre line of the ship and, when possible, within reach of the walkway or the fore and aft lifeline.
6 Release devices should be located so that they can be operated in all weather conditions. Devices which cannot be reached because of heavy seas breaking on board are of little use.
7 If the construction of the ship permits, a 'high level' wire safety line could be rigged after the cargo is stowed and before the vessel sails. The wire

could run from the bridge front to a forward stump mast. It may then be possible for a seaman wearing a safety harness to loop a line from the harness over the wire and back to the harness. The slip hooks can then be released in a forward to aft sequence. The high level wire should be high enough to be clear of any jettisoning cargo and if the seaman loses his footing or is injured it may be possible to bring him to safety.

8 The liberal application of oil and grease to the appropriate parts of the securing arrangements will greatly reduce the risk of releasing devices becoming inoperable.

It must be stressed once again that jettisoning cargo is extremely hazardous and every precaution should be taken to prevent a shift of cargo.

Loading and securing of a heavy weight on deck

The cargoes which are carried on deck are so diverse in nature that each load must be considered individually. However, the following guidelines should be observed when using the ship's 'jumbo' derrick:

1 In addition to the stability requirements in Part I of the regulations the stability at each stage of the lift should be considered, e.g. calculate the angle of heel at the moment the weight lifts off the quay.

2 All personnel should be warned before the operation commences so that accidents are not caused by an unexpected list.

3 The vessel should be on even keel and upright before the operation commences to assist the handling of the equipment. All tanks should either be full or empty to avoid free surface effects. The double-bottoms should be full.

4 All lifting appliances and equipment should be inspected before use and all items should be adequate and satisfactory for the job. The relevant regulations should be observed and all the equipment being used must have the appropriate certificates. It is important to include the weight of the lifting tackle with the weight of the load when deciding if the SWL of the derrick will permit the load to be lifted. Never exceed the SWL.

5 All leads must be fair to avoid chafing the falls. When possible, especially with a non-patent jumbo derrick, the purchase fall should be led through a mast head cargo purchase lead block. This reduces the weight/stress on the derrick span tackle and often provides a better lead to the winch. Ensure that no slack turns are allowed to occur on the winch barrel.

6 Additional slewing guys are sometimes attached to the bottom purchase block to reduce the strain on the derrick head. All guys and ancillary equipment should be well clear of any obstruction in all working positions of the derrick.

7 Ensure that all the ancillary equipment is rigged in correct locations, i.e. as per the rigging plan.

8 Where appropriate, the winches should be put into double gear.

9 Avoid lowering the derrick more than is necessary as this increases the stress on the derrick.

10 Adequate dunnage should be laid so as to spread the weight of the cargo over as much of the underdeck framing as possible. If shore cranes are used a heavy lift can be stowed above a transverse bulkhead as the bulkhead can distribute the weight of the load effectively.

11 There should be a sufficient number of securing points on the deck. If necessary, additional eye plates should be welded at suitable locations.

12 Check that there are sufficient lifting points on the cargo. Boilers and heavy plant will probably have permanent lifting attachments clearly marked. Carefully inspect the attachments and ensure that the slings are secured only to those points.

13 Check that the weight marked on the load corresponds with the weight stated in the cargo documents.

14 Boilers usually arrive on the quay already stowed in cradles. All cylindrical objects should be stowed in cradles similar to those shown in Figure 8.3.

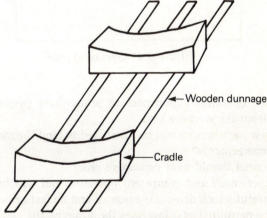

Wooden dunnage

Cradle

Figure 8.3   Deck cradle for a boiler

15 Cylindrical loads are often lifted by the use of eyes welded to the uppermost area. If slings are rigged around the body of a load, protection must be given to the load and to the slings to prevent any chafing or damage occurring.

16 With many heavy lifts special slings will be supplied and any slinging instructions should be carefully followed. The slings should never be

shortened and the eyes of slings should bear on the bow of the shackle and not the pin. Sling angles should never exceed 120°.

17 If a cargo hook is being used, it should be positioned above the centre of gravity of the load.

18 Some loads may be heavier at one end than at the other and may be of an irregular shape. In this case, slings with different leg lengths may be used and it is essential that each sling leg is attached to the appropriate securing position.

19 When possible use an 'I section spreader' or a beam sling to lift the load as these ensure an even distribution of weight during the loading operation.

20 Loads of an uneven weight distribution which are shipped in a box or container usually have the heavy end indicated by the 'seesaw' symbol shown in Figure 8.4. The centre of gravity should also be indicated.

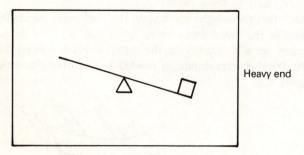

Figure 8.4   Seesaw symbol

21 Before the operation commences all unnecessary personnel should be cleared from the working area.

22 Some crew members should be instructed to tend the moorings and the access arrangements.

23 All personnel should wear protective gear.

24 Only experienced and competent operators and winchmen should be used. Careful winch driving is essential and no fast heaving or lowering should be permitted as it increases the strain on the equipment and is a dangerous procedure. All personnel should understand the pre-arranged signals. The load should not be jerked and must therefore be kept clear of bulwarks and hatch coamings.

25 The load should be kept as level as possible and steadying lines should be attached to the load where appropriate.

26 The cargo should be loaded as close as possible to hatch coamings or bulwarks so that the ship's permanent structure can be incorporated into the tomming-off and securing arrangements.

27 Wire rope lashings used for securing the load should be of the same strength as the wire ropes used for the loading operation. If necessary, obtain specialist advice from the shippers or from rigging companies as to the strength and type of securing arrangements necessary to keep the stow in place.

28 It may be possible to weld heavy items of machinery, such as power station turbine casings, to the ship's structure.

B   Grain Cargoes

Principle of the grain rules

The loading and transporting of grain is governed by Chapter VI of SOLAS 74, 'Carriage of Grain', generally referred to as the 'IMO Grain Rules'. The rules are relatively brief and should be read by students. The IMO rules have been formulated from international studies and the foreword to the rules explains that they are based on the fact that in a compartment filled with grain there exists a void space between the grain surface and the deckhead of the compartment. The basic requirement of the rules is that it must be demonstrated by calculation that at all times during a voyage the ship will have sufficient intact stability to provide adequate residual dynamic stability after taking into account the adverse heeling effects caused by an assumed pattern of grain movements within the void spaces lying directly above the grain. The foreword further states that the provision of temporary grain fittings to reduce the effects of grain shift, e.g. shifting boards, bagged grain, bundling, etc., 'depends entirely upon achieving the correct relationship between the intact stability characteristics of the ship and the heeling effects of a possible grain shift within the various compartments of the ship'. The IMO Grain Rules stipulate the minimum level of acceptable stability for the carriage of grain in terms of initial metacentric height, angle of heel due to assumed grain shift, and residual dynamic stability.

Merchant Shipping (Grain) Regulations 1980. SI 1985 No. 1217

The regulations apply to sea-going United Kingdom ships and to other ships while they are within the United Kingdom or territorial waters when loaded with bulk grain. The regulations give effect to Chapter VI of SOLAS 74.

Shipboard Operations

Where grain in bulk is loaded on board a British ship or is loaded on board any ship in a United Kingdom port, it must be loaded in accordance with the arrangements stated in the schedule to the regulations. Any ship entering the United Kingdom should have loaded any bulk grain cargo in accordance with the same regulations. A synopsis of the schedule is given below:

1 All free grain surfaces should be levelled to minimize the effect of the grain shifting. In any filled compartment the bulk grain must be trimmed so as to fill all the spaces under the decks and hatch covers. All free grain surfaces in partly filled compartments must be level.

2 Throughout the voyage the intact stability characteristics must meet at least the following criteria:
   (a) the angle of heel due to the shift of grain must not be greater than 12°;
   (b) in the statical stability diagram, the net or residual area between the heeling arm curve and the righting arm curve up to the angle of heel of maximum difference between the ordinates of the two curves shall, in all conditions of loading, be not less than 0.075 metre-radians;
   (c) the initial metacentric height, after correcting for the FSE of liquids in tanks, must not be less than 0.3 m. The ship must be upright before proceeding to sea.

3 In both filled and partly filled compartments, longitudinal or transverse divisions may be provided to reduce the adverse heeling effect of grain shift or to limit the depth of cargo used for securing the grain surfaces. Saucering may also be used to reduce the adverse heeling effects of grain shift.

4 The surface of bulk grain in any partly filled compartment should be secured by overstowing or by strapping or lashing.

The arrangements mentioned in 3 and 4 must be in accordance with the procedures specified in the IMO Grain Rules (see below).

Lower holds and tween deck spaces directly above may be loaded as one compartment provided that, in calculating transverse heeling moments, proper account is taken of the flow of grain into the lower spaces.

Document of Authorisation

M1253 draws attention to the fact that in accordance with the Merchant Shipping Grain Regulation; every ship which is to carry grain must have on board a 'Document of Authorisation'.

This document takes the form of a booklet of stability information which includes grain loading conditions and other information which indicates that the ship meets the relevant SOLAS requirements for carrying grain. Every booklet must be stamped by the Administration of the county in which the ship is registered, or by a Certifying Authority on its behalf. The stamp should bear the statement that the requirements can be met and it gives the authority for the ship to be loaded with bulk grain in accordance with those requirements.

Grain

The IMO definition of grain states that, 'the term "grain" includes wheat, maize (corn), oats, rye, barley, rice, pulses, seeds and processed forms thereof, whose behaviour is similar to that of grain in its natural state'.

Methods of securing grain

The Department of Transport has published a *Grain Notes Guidance Booklet* which indicates the type of information that should be provided with a carriage of grain submission. The booklet also describes the various methods by which grain is secured.

Longitudinal divisions

These may be provided in both filled compartments and partly filled compartments. The divisions must be grain-tight and constructed to IMO specifications.

In a filled compartment the division shall:

(a) in a tween deck extend from deck to deck;
(b) in a hold extend downwards from the underside of the deck or hatch covers a distance of at least 0.6 m below the grain surface after the surface has been assumed to shift through an angle of 15°.

In a partly filled compartment the division shall extend from one-eighth of the maximum breadth of the compartment above the level of the grain surface and to the same distance below the surface. When the division is used to limit the depth of overstowing, the height of the centreline division shall be at least 0.6 m above the grain level.

The specifications for grain fittings are given in Part C of the IMO rules. The rules should be consulted before divisions are constructed but some pertinent details are outlined below.

General specifications All timber used for grain fittings should be of good sound quality and of a type and grade which has been proved to be satisfactory for the purpose. Plywood, of an exterior type bonded with waterproof glue and fitted so that the direction of the grain in the face plies is perpendicular to the supporting uprights or binder, may be used provided that its strength is equivalent to that of solid timber of the appropriate scantlings. Materials other than wood or steel may be approved for divisions provided that proper regard has been paid to their mechanical properties.

Uprights   The depth of housing for each upright should not be less than 75 mm. This applies to the housing at both the upper and lower ends. If an upright is not secured at the top, the uppermost shore or stay should be fitted as near to the top as is practicable. If arrangements are made for inserting shifting boards by removing a part of the cross-section of an upright, the arrangements should be such that the local level of stresses is not unduly high. If uprights are formed by two separate sections, one fitted on each side of a division and interconnected by through bolts at adequate spacing, the effective section modules should be taken as the sum of the two moduli of the separate sections. If divisions do not extend to the full depth of the hold such divisions and their uprights should be supported or stayed so as to be as efficient as those which do extend to the full depth of the hold.

Divisions loaded on both sides

1 Shifting boards should be not less than 50 mm thick, grain-tight, and where necessary supported by uprights.
2 The maximum unsupported span for shifting boards of various thicknesses is given in a table, e.g. the maximum unsupported span for a 60 mm thick shifting board is 3.0 m.
3 The ends of all shifting boards should be securely housed with a 75 mm minimum bearing length.
4 Divisions formed with materials other than wood should have an equivalent strength to wooden shifting boards.
5 The section modules (cross-section size) of steel and wood uprights are determined by formulae which are given in the rules.
6 Uprights should be spaced to correspond with the maximum permitted unsupported span for the shifting boards.
7 If wood shores are used they must be in a single piece, securely fixed at each end, and heeled against the permanent structure of the ship (except directly against the side plating). The minimum size of the shores is given in the rules.
8 Stays must be as near horizontal as possible, well secured at both ends, and made of steel wire rope. The size of the wire should be determined on the basis that the divisions and uprights are uniformly loaded at 500 kg/m². The breaking load of the wire must exceed this by at least two-thirds.

The requirements for divisions which are loaded on one side only are found by the use of tables which are contained in Part C.

Saucering

Except in the case of compartments containing oil seeds, a longitudinal division beneath a hatchway in a filled compartment may be replaced by a saucer of bagged grain or other suitable cargo. The bulk grain beneath the hatchway should be trimmed in the form of a saucer and the following points should be observed:

1 The depth of the saucer, measured from the bottom of the saucer to the deck line, shall not be less than: 1.2 m in ships whose moulded beam does not exceed 9.1 m; and 1.8 m in ships whose moulded beam is 18.3 m or greater. In ships of moulded beam between 9.1 and 18.3 m the depth should be found by interpolation.
2 The top, or mouth, of the saucer should be formed by the hatchway, i.e. hatch side girders or coamings and hatch end beams. The saucer and the hatchway above should be completely filled with bagged grain or other suitable cargo which is laid down on a separation cloth or its equivalent. The bags must be stowed tightly against adjacent structures and portable hatchway beams if the latter are in place.

Bundling

As an alternative to filling the saucer with bagged grain or other suitable cargo a bundle of bulk grain may be used. The grain is trimmed in the form of a saucer as described above and the following procedure should then be carried out:

1 The saucer should be lined with a tarpaulin or other acceptable material which has a tensile strength of not less than 274 kg per 5 cm strip. The material must be provided with suitable means for securing at the top, e.g. strong eyelets and suitable lashings.
2 The saucer should then be filled with bulk grain and the top secured, usually carried out by lacing the eyelets together and forming a tight compact bundle. The top of the saucer must be coincidental with the bottom of the beams when they are in place and suitable general cargo or bulk grain may be placed between the beams on top of the saucer to ensure a tight, compact stow.

If more than one sheet of material is used to line the saucer the sheets must be joined at the bottom by sewing or by a double lap.

Acceptable material of a strength less than 274 kg per 5 cm strip may be used but it must not have a tensile strength of less than 137 kg per 5 cm strip. If such material is used, athwartship lashings must be placed inside the saucer directly on top of the grain at intervals of not more than 2.4 m. The lashings must be long enough to be drawn up tight and secured at the top of the saucer. Dunnage which must not be less than 25 mm thick (or other suitable material of equal strength) and between 150 and 300 mm wide should then be placed fore and aft over the lashings. The purpose of the dunnage is to prevent the lashings cutting or chafing the material which is placed on top of the dunnage to line the saucer. The bundle is then drawn up tight and secured as previously described.

Saucering and bundling can only be carried out in filled compartments and the procedures are similar, except:

> for saucering bagged grain must be used
> in bundling bulk grain is used

In some ports the trimming of grain to form a saucer is done manually and the personnel conducting the operation should be made aware of the dangers of oxygen deficiency which may be caused by the grain 'breathing'. All organic materials consume oxygen and some grains can quickly use up the oxygen in confined spaces. 'Enclosed space procedures' should be adapted for the trimming operation and, if necessary, the personnel should wear some form of air supply apparatus. Accidents have occurred in which men trimming grain have collapsed and died as a result of oxygen deficiency.

Partly filled compartments

The movement of grain in a partly filled compartment may be considered to be eliminated if the grain is overstowed or strapped.

Overstowing
1 The surface of the grain must be trimmed level.
2 A separation cloth (or equivalent) or a platform is laid on top of the grain. A platform consists of wooden bearers spaced not more than 1.2 m apart with 25 mm wooden boards laid on top not more than 100 mm apart.
3 Bagged grain is tightly stowed on top to a height of not less than one-sixteenth of the breadth of the free grain surface or 1.2 m, whichever is the greater. (Instead of bagged grain, other suitable cargo which exerts the same pressure may be used.) The bagged grain must be in securely closed, well filled, sound bags.

Strapping or lashing
1 The grain is trimmed so that the surface is very slightly 'crowned', i.e. the surface is slightly higher in the middle than around the edges.
2 The grain is covered by separation cloths and/or tarpaulins which overlap by at least 1.8 m.

3 Two solid floors of 25 mm timber, 150–300 mm wide, are laid on the separation. The top floor runs longitudinally and is nailed to the athwartships bottom floor. Alternative approved floors may also be laid.
4 Timber bearers, 25 × 150 mm, are nailed to the top floor. The bearers extend over the full breadth of the compartment.
5 The floors are secured by lashings spaced not more than 2.4 m apart and which lie on the bearers. The lashings are made of 19 mm diameter steel wire rope, or 50 × 1.3 mm doubled steel strapping (having a breaking load of at least 5000 kg), or chain of equivalent strength. The lashings are set tight by 32 mm turnbuckles but winch tighteners with locking arms may be used with steel strapping. Eyes in lashing wire must be formed by at least four clips on each eye and the ends of steel strapping must be secured by not less than three crimp seals on each end.
6 The lashings will have been placed in position prior to the completion of loading by being positively attached to framing at points approximately 450 mm below the anticipated final grain surface by means of either a 25 mm shackle or a beam clamp of equivalent strength.
7 The lashings must be regularly inspected during the voyage and re-set where necessary.

C Solid Bulk Cargoes

Solid bulk cargoes should be loaded and transported with extreme care and should be considered to be dangerous cargoes. Improper transportation of bulk cargo is a major factor in ship losses. The size of the ship would appear to have little relationship to the hazards of transporation as losses have been in the range between 1000 and 115,000 grt. It is therefore essential to observe strictly recommendations which pertain to the safe carriage of solid bulk cargoes.

Code of Safe Practice for Solid Bulk Cargoes

This excellent code is published by IMO and the sixth edition was published in 1987. The code should be cross-referenced with supplements which are issued as circumstances require them; the first supplement to this edition was published in 1988.

The code contains an introduction, ten sections, six appendices, an index of cargoes and a removable Maritime Safety Card on enclosed space procedures:

Introduction—The aim of the code
Section   1—Definitions

All the sections are concise and practical and should be carefully studied in full by all persons engaged in, or associated with, the transportation of solid bulk cargoes by sea.

Introduction

This states that the primary aims of the code are to highlight the dangers of certain types of bulk cargo, to give procedural guidance, to list typical products together with advice on their properties and on handling, and to describe test procedures for determining cargo characteristics.

The main hazards of shipping bulk cargoes are:

1 Structural damage due to improper distribution of the cargo
2 Loss or reduction of stability from
   (a) a shift of cargo
   (b) cargoes liquefying
3 Chemical reactions

Attention is drawn to the relevant sections which deal with the above hazards.

Section 1—Definitions

There are fifteen definitions in the section and mariners should be conversant with them all. Six definitions are of particular importance:

Angle of repose   The angle between a horizontal plane and the cone slope of a cargo.

Flow moisture point The percentage moisture content (on a wet mass basis) at which a flow state develops under the prescribed method of test in a representative sample of the materials.

Flow state The state that occurs when a mass of granular material is saturated with liquid to an extent that under the influence of prevailing external forces such as vibration, impaction, or the ship's motion, it loses its internal shear strength and behaves as a liquid.

Moisture content The portion of a representative sample of a material which consists of water, ice, or other liquid expressed as a percentage of the total wet mass of that sample.

Moisture migration The movement of moisture contained in a bulk cargo by settling and consolidation of the cargo due to vibration and ship's motion. Water is progressively displaced which may result in some portions or all of the bulk cargo developing a flow state.

Transportable moisture limit (tml) The transportable moisture limit of a cargo which may liquefy represents the maximum moisture content of that cargo which is considered safe for carriage in ships which are not specially fitted or constructed for cargoes of excessive moisture content. (The tml is 90 percent of the flow moisture point.)

Section 2—General precautions

Bulk cargoes must be properly distributed throughout the ship in order that the structure will never be overstressed. Detailed information should be provided for the loading of high-density cargoes but if such information is not available then the following precautions should be observed:

1. The general fore and aft distribution of weight should not differ appreciably from that found satisfactory for general cargoes.
2. The maximum number of tonnes of cargo loaded in any space should not exceed 0.9LBD tonnes

    where L = length of the hold in metres

       B = average breadth of the hold in metres

       D = the summer load draught in metres
3. If cargo is untrimmed, the height of the cargo pile above the floor should not exceed in metres

       1.1 × D × stowage factor

    where the stowage factor is given in cubic metres per tonne.

4 Other guidelines apply to cargo which is trimmed entirely level and to holds divided by shaft tunnels.

The stability information booklet should be studied and the ship's stability calculated for the anticipated worst conditions during the voyage. In general, high-density cargoes should normally be loaded in lower holds rather than in tween decks but an excessively high GM should be avoided as it can produce violent movement during the passage.

Section 3—Safety of personnel

Particular attention is drawn to poisoning and asphyxiation hazards, health hazards due to dust, and the dangers of flammable atmospheres.

Section 4—Assessing the acceptability of consignments

Before loading commences the Master should be provided with appropriate information concerning the characteristics and properties of the cargo. Such information should contain as a minimum:

> chemical hazards
> flow moisture point
> stowage factor
> moisture content
> angle of repose

The Master should be given a certificate or certificates containing the relevant characteristics of the cargo. The certificates should include:

> a certificate which states the transportable moisture limit
> a certificate of moisture content
> a statement that the moisture content specified in the certificate of moisture content is the actual moisture content of the cargo at the time the certificate is given to the master

If materials possess chemical hazards, a certificate listing the hazards should be given to the master. The certificate should contain a statement that the chemical characteristics are those existing at the time of loading.

The section also gives information on sampling procedures and lists the factors which must be taken into account when obtaining representative samples:

1 Type of material.
2 Particle size distribution.
3 Composition of the material and its variability.
4 Manner in which the material is stored, e.g. in stockpiles or railway wagons, and the method by which the material is transferred or loaded,

e.g. by conveyors or grabs.

5 Characteristics which must be determined, e.g. moisture content and flow moisture point.

6 Weather conditions which may cause moisture distribution variations throughout the consignment.

Section 5—Trimming procedures

For the purposes of trimming, bulk cargoes are placed into two general categories:

1 *Cargoes with an angle of repose less than or equal to 35°* Mariners can consider this category to be the more dangerous of the two as cargoes of this type are subject to the liquefaction process which will be discussed later. Cargoes of low angle of repose generally consist of small granules which shift quite easily when the vessel moves. Small granules tend to move over one another easily and thus if the vessel rolls the cargo will shift to the low side. Most small granules absorb moisture and if the weatherdeck hatch leaks the cargo will absorb water entering the hold. Thus the hold may eventually contain a cargo of sludge which, of course, is detrimental to the ship's stability.

This section recommends that cargoes of such an angle of repose be trimmed reasonably level and that the spaces in which the cargoes are loaded should be filled as fully as practicable without resulting in an excessive cargo weight in one location. If cargoes which flow like grain are to be carried, the stowage procedures for grain should be observed. However, the density of the cargo should be considered when determining:

(a) the construction of securing arrangements; and

(b) the stability effect of free cargo surfaces.

2 *Cargoes with an angle of respose greater than 35°* These cargoes often consist of lumps of various sizes, e.g. from 3 to 300 mm, which interlock and are therefore quite stable. The cargo can become heaped up and produce an excessive pile peak. A major hazard with this type of cargo is that if the cargo is left untrimmed and the vessel rolls at sea, the cargo may fall into the unoccupied cargo space on one side of the hold and, as the cargo is relatively stable, remain there. This, of course would produce a list.

This type of cargo is not usually subject to liquefaction as any liquid entering the hold remains in the unoccupied space between the lumps and eventually filters through to the bilges, from where the ship's bilge pumps can remove the liquid.

The section recommends that when cargo is loaded only in lower cargo spaces, it should be trimmed sufficiently to cover all of the tank top out to the ship's side in order to reduce the pile peak and to equalize the distribution of weight. If it is necessary to load cargo in a tween deck, the tween deck hatch

261

should be closed. The tween deck cargo should be trimmed reasonably level and should either extend from side to side and bulkhead to bulkhead or be secured in bins.

Section 6—Methods of determining the angle of repose

Attention is drawn to laboratory, shipboard, and on-site methods. The on-site method requires the angle subtended by the surface of a stockpile with the ground to be measured at a minimum of six points around the circumference of the stockpile and the mean taken. If it is not possible to examine a stockpile, a cone of material deposited on to the quay by a grab could be measured in a similar way. However, the larger the cone is, the more accurately the angle of repose will be to that produced in the ship's hold.

Section 7—Cargoes which may liquefy

This is an extremely important section and all officers involved in carrying bulk cargoes should be well conversant with the contents.

Cargoes which may liquefy include concentrates, certain coals, and other materials having similar physical properties. Section 1 defines concentrates as materials which are obtained from a natural ore by a process of purification by physical or chemical separation and removal of unwanted constituents. Appendix A contains a list of cargoes which may liquefy. Such cargoes usually consist of small particles of uniform size (often resembling dust) in contrast to natural ores which include a considerable percentage of large particles or lumps. Concentrates are heavy 'dense' cargoes and the stowage factor is very low, generally between 0.33 and 0.57 m^3/t. If there is any doubt as to the nature of a bulk cargo, it should be considered as a cargo which is liable to liquefaction.

Concentrates have a property which is similar to that of blotting paper or a sponge. Drop some water on to one corner of a sponge and it will absorb moisture. Keep dropping water on to the same portion of the sponge and moisture will eventually permeate through the material until all the sponge is wet. A similar process can happen to a hold full of a concentrate. If one corner of a hatch cover leaks and a steady flow of moisture enters one corner of a hold, the moisture will not remain in that area. The cargo will absorb the moisture until all the concentrate in the hold is wet. If liquid permeates through a sufficient amount of cargo a flow state develops in the cargo. This liquefaction produces a very dense sludge which moves about the hold and, in addition to the problem of free surface effect, the dense nature of the cargo may cause structural damage. If a list develops there is a real chance that a negative GM will develop and the ship may eventually be lost.

In technical terms, if the cargo has, or attains, a moisture content above that of the transportable moisture limit, shift of cargo may occur as a result of liquefaction. Cargoes of this nature may appear to be in a relatively dry granular state when loaded. However, the cargo may contain moisture which cannot be detected visually and under the stimulus of compaction and vibration which occurs during a voyage the cargo my become fluid. Section 7 points out that the viscous cargo fluid may flow to one side of the ship with a roll one way but not completely return with a roll the other way. The ship may progressively reach a dangerous heel and capsize.

The following general precautions should be observed:

1 These cargoes should be trimmed reasonably level on completion of loading irrespective of the angle of repose.
2 The moisture content should not be in excess of the transportable moisture limit.
3 Cargoes which contain liquids should not be stowed in the same cargo space above or adjacent to these cargoes.
4 No liquid should be permitted to enter the cargo spaces (contact with sea water could also lead to serious corrosion problems, e.g. sulphur + water = sulphuric acid).
5 If it is necessary to cool the cargo, the water should be applied in the form of a spray to reduce the possibility of inducing a flow state.

Section 8—Test procedures for cargoes which may liquefy

This section gives an auxiliary test check which may be carried out on board ship if the ship's officers have doubts concerning the condition of cargo for safe shipment. The test approximately determines the possibility of flow.

A cylindrical can or similar container of 0.5−1 litre capacity should be half-filled with a sample of cargo. The can should be held in one hand about 0.2 m above a hard surface. The surface should be struck 25 times at one or two second intervals and the material examined. If free moisture appears or a fluid condition exists, arrangements should be made to have additional laboratory tests carried out on the cargo before it is accepted for loading. It must be emphasized that this test does *not* indicate that the cargo is safe for shipment; it only indicates that the cargo may be unsafe.

Section 9—Materials possessing chemical hazards

The hazards are classed and described in the manner laid out in the IMDG code. However, attention is particularly drawn to 'Materials hazardous only in bulk (MHB)' as these materials may not be listed in the IMDG code, being only hazardous when carried in bulk. For example, some cargoes are liable to reduce the oxygen content in a cargo space and other cargoes are liable to self-heating. Cargo segregation information is also given in this section.

M notices

M899 Stowage of steel, ores, or other heavy cargoes in tween decks

This notice draws attention to the fact that several cases have occurred in which ships' tween decks and hatches have collapsed due to loading of cargo much in excess of that for which decks are normally designed.

Heavy cargo should, therefore, be so distributed that the intensity of loading on any part of a tween deck or hatch does not substantially exceed the weight of cargo that could be stowed thereon for the full tween deck height at the rate of 1.39 m^3/tonne.

In a 2.44 m tween deck the advisable height for, e.g. bulk ore stowing at 0.53 m^3/tonne would be

$$2.44 \times \frac{0.53}{1.39} = 0.93 \text{ m}$$

M1250 Precautions against liquefaction of coal slurry, duffs, small coals and coke

The principal aims of this notice are to draw attention to the dangers associated with types of coal cargoes which can in certain circumstances liquefy, and to provide advice on precautions to reduce the risk of such dangers.

The notice should be read in full as it contains much valuable information which could be related to the shipping of all cargoes which have a potential for liquefaction. The notice should be read in conjunction with M1246 and both notices contain some similar and complementary information.

M1246 The safe carriage of solid bulk cargoes

The notice draws attention to the IMO code and states that the provisions of the code are strongly recommended by the Department of Transport. However, the principal aim of this notice is to supplement the code in respect of cargoes which might liquefy. Once again, this notice should be read in full as it contains much excellent, practical information.

The notice points out that there are three principal sources of moisture in bulk cargoes:

1 Water associated with the mining of the raw materials and water resulting from the manufacturing or preparation process.
2 Exposure to rain, snow, and ice.
3 Stockpiling on wet ground.

The notice also contains information on sampling, testing, precautions

264

prior to, during, and after loading, and the provision of cargo information.

A combined form of certificate and statement which will satisfy IMO requirements is shown in Appendix 1. The form sets out:

1 The flow moisture point and the transportable moisture limit.
2 The average moisture content of the complete cargo or the moisture content of the cargo in each hold.
3 The angle of repose.
4 Any other relevant information on the physical properties of the cargo.

Appendix 2 contains guidance on the arrangements for sampling from wagons, lorries, and stockpiles.

Appendix 3 gives advice on 'The Speedy Moisture Tester' which is an instrument for providing a quick, direct reading method of ascertaining the free moisture content of a granular product.

The information provided by the code and the M notices should be collated and combined with practical experience to enable the mariner to deal with particular operational problems. Three aspects of the transporting of bulk cargoes are considered below.

Preparations before loading bulk cargoes

The cargo spaces should be prepared for the particular bulk cargo which is to be loaded, e.g. detachable spar ceiling should be removed from lower holds and stowed in suitable tween deck locations. All weatherdeck hatches should be hose tested and any leaks should be made good. All hatch-securing arrangements should be overhauled. The holds must be dry and all bilge suctions, pipes, and pumps should be tested. All sounding pipes, air pipes, ventilators and service pipes within the cargo compartments should be inspected to ensure that they are in good order. It may be necessary to place additional strainer plates, or fit temporary straining cloths, over bilges wells to prevent cargo material entering the bilge pumping system.

Precautions should be taken to prevent dust entering deck machinery, the hold ventilation system if fitted, engine room air inlets, and the ship's electronic navigation equipment. It may be necessary to shut down the air-conditioning system, and all electrical equipment used during the cargo operation should be incapable of causing a dust explosion.

All appropriate fire precautions should be taken and officers should ensure that the ship's fire-fighting system is suitable for the intended cargo.

The IMO code should be consulted to obtain the physical properties, segregation and stowage requirements, health hazards, and any special requirements pertaining to the cargo. If necessary, *Medical First Aid Guide for Use in Accidents Involving Dangerous Goods* (MFAG) should be consulted to ensure that the vessel has the proper medical facilities should an incident occur.

It is possible that details of the cargo to be loaded will not be found in the IMO code. The shipper must therefore provide all the necessary information to enable the cargo to be loaded properly. It may be necessary to communicate with the authorities at the discharging port to check if the cargo can be discharged there.

On occasions, attempts have been made to persuade Masters that the IMO certification procedures are unnecessary. In such cases the owners should be notified and pressure brought to bear on the shippers to produce the appropriate documents. The Master may have to take a firm stand and refuse to allow loading to commence until test certificates have been produced.

The ship's staff should try to ascertain whether the samples have been taken in accordance with the guidelines in the IMO code and M notices. Appendix 2 of M1246 should provide the basis of the sampling technique, with due allowance being made for local circumstances. If the cargo is to be loaded from wagons, a kilogram of material should be taken at one-third of the depth of each wagon. When 10 kg have been assembled in a dry place the sample should be tested. Where samples are taken from a stockpile, the samples must be representative of the complete stockpile.

Officers should inspect the cargo material before samples are taken and should reject any consignment which is unsuitable for shipment. The shipboard tests for moisture content which are specified in M1246 should be carried out. An additional test that is used on some vessels transporting bulk cargoes appears to have practical value. Cargo samples are heaped on a tray which is placed near the ship's generators. If the sample contains excessive moisture the vibration of the generators will cause liquefaction to occur. As with the cylinder test, an absence of moisture will not indicate that the cargo is safe. However, the generator test may indicate that something is decidedly wrong. The tests and procedures should not be carried out only on concentrates but on all cargoes of a similar nature. In addition to the problem of liquefaction, moisture may cause chemical hazards in certain cargoes.

Loading and carriage of bulk cargo with relation to stress in a general cargo vessel

The mariner must consider the effect of cargo distribution on local areas as well as on the ship's overall structure. The general fore and aft distribution of weight should, therefore, closely follow that for any general cargo which has been previously carried.

To avoid local stress the maximum weight in any one space should not exceed 0.9LBD tonnes and the height of the cargo pile peak should not be excessive. The incorrect loading of a vessel can set up shear stresses between adjoining compartments, especially if one compartment is full and the other empty. A pre-calculated loading sequence must be adhered to in which one hold is only partly loaded when other holds are still empty. This is particularly important at berths which have only one loading point and the vessel is moved up and down the berth by her moorings to load holds in a sequence.

The overall distribution of cargo should be such as to avoid hogging and sagging stresses. These stresses should be pre-calculated for the loading and discharging operations as well as for the voyage. Vessels have been subjected to overall structural damage which has been caused by incorrect cargo-handling operations. Difficulty may be encountered when securing cargo hatches on ships which hog or sag and the water-tight integrity of the ship may be impaired.

An excessively high GM should be avoided and it may be necessary to load cargo into tween deck spaces in order to reduce the ship's metacentric height. The recommendation in M899 (heavy cargoes in tween decks) should be observed so as to avoid overstressing the tween deck structure. A ship with a large GM will also have a large righting lever and thus a stiff ship is liable to roll violently in a seaway. The short roll period and 'whiplash' motion could cause racking stresses and the transverse members of the ship, e.g. transverse bulkheads, beam knees and plate floors, could be damaged by such stress. The violent rolling could also cause cargo to shift and structural damage could result from cargo movement stress.

Structural damage could also be the result of stress caused by incorrect loading techniques. Double-bottoms should be pressed up. This will provide temporary 'stiffening' and support for the inner bottom structure and will also help to disperse energy shocks from the impact stress of cargo landing in the hold. If cargo is loaded by grabs, the first few loads should be lowered into the hold and deposited from a height of a few metres above the inner bottom, not from the height of the weatherdeck coaming. If the bulk is loaded from shutes or conveyor belts the initial rate should be slow. Some ships have found it a wise precaution to leave any portable tween deck hatch cover beams in position to provide stiffening for the hatch structure. When possible, the cargo should be trimmed during the loading process to avoid excessive pile peaks causing stress on the double-bottoms.

If ballast is to be pumped out during loading the stress caused by the concurrent de-ballasting/loading operation must be calculated carefully. Structural damage has been caused in port by incorrect ballast/cargo weight distribution.

Inherent hazards of concentrates

The nature of concentrates should first be considered. Concentrates are obtained from natural ore by a process of purification which removes unwanted constituents. They therefore usually consist of pure mineral particles of a dense nature which flow like fine sand. The problems of moisture migration producing a flow state have already been discussed but mariners must never underestimate the effect which liquefaction will have on a ship's stability. Even without moisture entering a cargo compartment the low angle of repose of concentrates makes a shift of dry cargo likely unless measures have been taken to secure the cargo.

A study of casualty reports shows that liquefaction is probably the major single hazard associated with the carriage of bulk concentrates. However, other hazards, which come under the general term of chemical hazards, must be considered:

1 *Fire hazard* Some concentrates are subject to spontaneous combustion (i.e. they set themselves on fire if the temperature is high enough) and others will ignite if an external heat source is applied to them.
2 *Explosion* The dust of some cargoes can form an explosive atmosphere when mixed with air, e.g. sulphur.
3 *Suffocating gas* Some concentrates evolve suffocating gas when involved in a fire, e.g. sulphur.
4 *Toxic gas* Many concentrates produce toxic gases in various ways. Ammonium nitrate fertilizers produce toxic gases upon decompositon.
5 *Dust inhalation* The dust of most concentrates should be avoided as many are toxic if the dust is inhaled, e.g. antimony ore.
6 *Swallowing* Personnel are unlikely to swallow cargo directly but dust can settle on food and many concentrates are toxic if swallowed, e.g. barium nitrate.
7 *Contact* Direct contact with the cargo should be avoided as even seemingly innocuous cargoes can cause health problems. Castor beans when handled can give rise to severe irritation of the skin and eyes.
8 *Oxidation* This is the process of combining with oxygen and it usually produces a chemical reaction. Some concentrates produce heat during oxidation (e.g. sulphides) and precautions may have to be taken to reduce contact with air. Oxygen reduction occurs with oxidation and 'Enclosed space entry procedures' should be observed before entering any compartment which contains concentrates.
9 *Water contact* Liquefaction is not the only problem caused by water. Aluminium dross may heat and evolve flammable and toxic gases upon contact with water and other concentrates have similar properties.
10 *Corrosion* Some concentrates cause corrosion because of galvanic action and others can mix with water to produce corrosive acids, e.g. sulphur.

Thus all concentrates should be considered to be hazardous cargoes and few are innocuous. Accidents have even occurred when concentrates have made the decks slippery. Thus minor hazards have brought home the point: *never trust concentrates.*

Further reading

Journal

CAINES, C. 'Changes in Stowage Factors when Frozen Bulk Cargoes Thaw' *Seaways*, November 1986, pp 3–4.

FOY, D. and SHEARER, I. *'Berge Vanga*—The Inevitable Disaster?' *Seaways*, September 1980, pp 5–6.

FOY, D. 'Bulk Carrier Losses—Unanswered Questions' *Seaways*, May 1988, pp 21–22.

Magazine

'Loss of the *Pool Fisher*', *Safety at Sea*, May 1981, page 29.

McCURDY, J. W. 'The Death of Thirty-One Men', *Safety at Sea*, November 1981, pp 34–37.

FOY, D. 'Combination carriers, ore carriers and sudden sinkings', *Safety at Sea*, February 1983, pp 29–33.

Monograph

KNOTT, J. R. *Lashing and Securing of Deck Cargoes* (The Nautical Institute, 1985).

9

Derricks

Some mariners have the misconception that derricks are 'old-fashioned' and are an out-of-date cargo-handling method. However, derricks would still appear to be the fastest and most efficient method of handling general cargo. A study of the 1980 edition of *Standard Ship Designs—Dry Cargo Vessels 9500 – 24,000 dwt*, which is produced by Fairplay Publications, shows that approximately 64 percent of the designs are for all derrick ships, 18 percent are for all crane ships, and 18 percent for mixed cranes and derricks.

The designs are the work of forty-six principal shipbuilders and would indicate the belief within the shipping industry that derricks are the most versatile of the ship cargo-handling methods. A typical design is that of a 17,200 dwt dry cargo carrier from Harland and Wolff of Belfast (Figure 9.1).

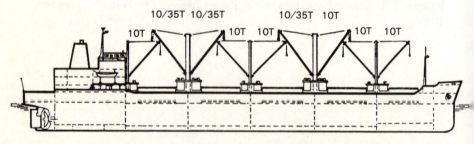

Figure 9.1   Design for a 17,200 dwt dry cargo carrier (Harland and Wolff, Belfast)

This design has eleven 10T derricks plus three 10/35T derricks, and a heavy lift version is available with a 100T Stuelcken derrick. The design of a 16,500 dwt cargo vessel from Sunderland Shipbuilders shows that the cargo gear consists of ten 10T derricks and one 50T derrick. Thus, it is apparent that ship's officers will be handling derricks for the forseeable future.

The fitting and operation of ships' cargo-handling appliances are strictly governed by the British Standards Institute and by Classification Societies.

Personnel involved with cargo handling should be acquainted with the following documents:

'Design and operation of ships' derrick rigs', BS MA 48: 1976
'Ships' derrick fittings', BS MA 81: 1980
'Jib cranes: ship mounted type', BS MA 79: 1978
'Code for Lifting Appliances in a Marine Environment', Lloyd's Register of Shipping

Union purchase system

This is probably the most common derrick system in use on general cargo vessels (Figure 9.2). Two derricks are 'coupled', 'married', or joined by a

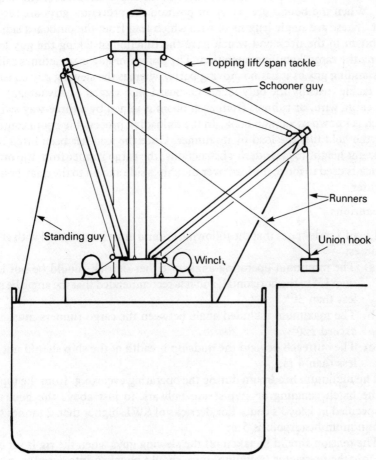

Figure 9.2   Union purchase rig (slewing guys not shown)

271

union hook and are worked in conjunction with each other. Each cargo boom is joined to the vertical mast or post by a swivel fitting known as a goose neck (so named because of the shape of the fitting). The up and down, or luffing, movement of the boom is carried out by a topping lift/span tackle, and the horizontal or athwartships movement is controlled by a slewing guy attached to the outboard side of the boom head. The two booms are linked by a schooner guy which runs from the inboard side of one boom head to the other and thence to the deck via a lead block on the mast. The schooner guy is sometimes replaced by inboard slewing guys but the latter tend to interfere with the cargo-working operation. The schooner guy is always well clear of the cargo working area.

The derricks are positioned by the guys and tackles. One boom is positioned over the hatch and the other boom is positioned over the ship's side. When the booms are set up in position the preventer guys are set up tight. These are single lengths of wire which lead from the outboard side of the boom to the deck and which have the function of taking the guy load during the cargo-handling operation. The preventer guy is sometimes called the standing guy as it has no moving parts, whereas the slewing guy consists of a tackle (usually the only tackle on board ship rigged to advantage).

A cargo wire, or runner, from each boom is joined by a three-way swivel which is known as a union hook. In the unloading process the boom centred over the hold lifts the load by its runner. Once the load has been lifted to a sufficient height to clear deck obstructions, the cargo runner from the other derrick is used to move the load over the ship's side and on to the quay or into a lighter.

Precautions

1 Lloyd's code states that the following criteria must be complied with at all times:
   (a) The minimum operating angle of either derrick should be not less than 15° to the horizontal, and it is recommended that the angle be not less than 30°;
   (b) The maximum included angle between the cargo runners must not exceed 120°;
   (c) The outreach beyond the midship breadth of the ship should not be less than 4 m.
2 The minimum headroom during the operating cycle (e.g. from the top of the hatch coaming or ship's side bulwark to just above the hook) is specified in Lloyd's code. For derricks of SWL higher than 2 tonnes the minimum headroom is 5 m.
3 The tension should be taken off the slewing guys when the rig is set up. Only the preventor (standing) guys should be taken into account in the calculation of forces in the rig.

4 BS MA 48: 1976 states that 'the practice of using an arbitrary percentage of the SWL to derive a value of SWL is not only unscientific but it can be misleading'. However, the Code of Safe Working Practices for Merchant Seamen recommends that 'where derricks have not been marked with the safe working load in union purchase they should not be used for loads in excess of one-third of the SWL of the derrick'. Load diagrams for the ship's union purchase rig should be consulted and if necessary the section in BS MA 48: 1976 relating to 'Methods of calculating loads in union purchase rig' should also be consulted. The maximum SWL of the rig should never exceed the SWL of the cargo runner and a good safety factor would be to lift a maximum weight equal to 75 percent of the runner SWL.

5 Unduly long slings should not be used by stevedores.

6 The standing guy of the boom which is over the side should not be placed too far back as this will increase the possibility of the boom 'jacknifing'.

7 The same guy should not be secured too close to the boom as a narrow angle will increase the loading on the guy.

8 Winch operators should keep the headroom beneath a load to the minimum required for safety. This keeps the forces within the rig to a minimum.

9 The standing guys must be of adequate strength. A standing guy should not be secured to the same eye plate on the spider band (a band of eye plates at the head of the derrick) to which the slewing guy is secured. The deck eye plate to which it is secured should be in an appropriate location.

10 A union hook should be used to connect the cargo runners. The three-way swivel, i.e. there is a swivel above the cargo hook and one at each runner shackle, will ensure that twists will not develop in the runners.

11 Runners should not be permitted to rub against hatch coamings or other structural members as this practice will quickly render them unfit for service.

12 The rigging of the derricks should be supervised by a competent person.

13 Protective equipment should be worn by persons involved in the operation.

The main advantage of this system is that it is probably the fastest method used for discharging break-bulk, non-unitized general cargo.

Disadvantages

1 It can only be used for light loads, an average of approximately 1.5–2 tonnes per load.

2 The winchmen must be highly skilled and experienced.

3 The derricks cannot be used for 'spot loading'.

4 Re-positioning the derricks is time-consuming.

Wire sizes

There are no 'standard sizes' for the wire ropes used in the union purchase system. However, Lloyd's code states that 'wires ropes for running rigging are to be constructed of not less than six strands over a main core'. In general, each strand must consist of not less than nineteen wires and may have a fibre or a wire core. Wire ropes for standing rigging are generally to be constructed of six strands over a wire core. Wire size and strength must correspond to the SWL of the derrick and attachments but the following can be taken as an indication of wire rope size for a union purchase rig:

Slewing guys	20mm diameter	6 × 19, strand construction	12 + 6 + 1,	fibre core
Standing guys	24mm	6 × 19	9 + 9 + 1,	wire core
Topping lift	24mm	6 × 24	15 + 9	, fibre core
Cargo runner	18mm	6 × 24	15 + 9	, fibre core

It would be inappropriate to give wire lengths but the average length of a runner is approximately 70 m.

On some vessels the union purchase has been developed into the mechanized coupled derrick system. This system generally uses six winches, i.e. two topping lift winches, two slewing winches, and two cargo winches, although the use of split drum winches can reduce the total number of winches. The topping lifts usually run to independent remote-control winches, and in conjunction with remote-control guy winches, mechanize the resetting of the two coupled derricks. The repositioning of the derricks can therefore be carried out in one minute by one man working at a central control position. The system is more expensive to install than the basic union purchase system but port operational costs are reduced.

Patent derricks

Several patent derrick systems have evolved from the basic swinging or slewing derrick. Most patent derricks have similar basic characteristics:

1 The twin topping lift/slewing guy principle is used which gives good control of a single derrick.
2 The capability of handling heavier loads than the union purchase system.
3 Combined slewing and topping (luffing) tackles.
4 Very good spot loading facilities, i.e. the load can be set down in most positions within the hatch area.
5 A high degree of centralized control with the operation being conducted by one man.
6 The derrick is rigged at all times and can quickly be brought into operation.

7 The use of new technology reduces the stresses encountered with the union purchase system.

There are many patent derrick systems used on board ship but the best known are probably 'Hallen' and 'Velle' for the handling of general cargoes, and 'Stuelcken' for heavy lifts.

Hallen derrick

The Hallen swinging derrick employs the twin topping principle which allows good control of a single derrick. This derrick was originally designed for loads of 5–8 tonnes but loads of over 100 tones are now unexceptional. The derrick can be mounted on all types of mast or derrick post and can make a traverse from port to starboard of 160–180°.

In the original design a fixed frame 'outrigger' was fitted to the mast (Figure 9.3) which was commonly known as a 'D' frame. This had the effect of keeping the topping lifts at a sufficiently wide angle to one another to ensure the derrick remaining steady even when swung out over the ship's side to an angle of 80° from the fore and aft line. The D frame also helped to keep the derrick stable in all positions, even when the vessel had a list. However, under some operational conditions there were disadvantages when using the D frame:

1 When the derrick was swung outboard, the sharp angle created by the contact of the topping life guy pennant with the frame caused excessive strain in the topping lift.
2 There was a tendency for the single-wire pennant on the topping lift to slip above or below the frame when working at 'difficult' angles, once again putting excessive strain on the topping lift.
3 The contact with the frame caused chafing on the pennant. This was reduced by fitting rollers to the frame or by protecting the wire.

The D frame has been largely replaced by outrigger rods (Figure 9.4) which are pivoted, and are stayed on the outboard side only so that the rod nearest the discharging side can swing towards the ship's side, thus ensuring a wide separation angle of the topping lifts.

As with other patent derricks, such as Velle and Stuelcken, the V-shape arrangement of the topping lifts gives a broad base which is necessary for lateral holding and guiding of the derrick. In Figure 9.4 the broad base between the topping lifts is provided by a cross-tree at the mast head. It could also be provided by derrick posts, gate masts, or V masts.

In the Hallen system each topping lift runs to its own winch. Hauling on both winches tops the derrick, and if one winch hauls in while the others pay out, the derrick slews to the side of the ship on which the hauling winch is located. A third winch is used for hoisting and lowering the cargo. The

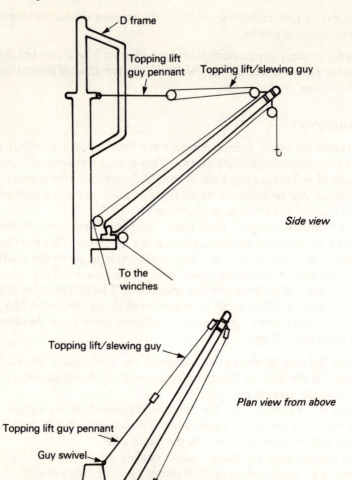

Figure 9.3   Hallen 'D' frame

derrick is controlled by two levers. One lever operates the cargo purchase and the other lever has a multi-position control for the topping and slewing operation.

Velle derrick

The Velle swinging derrick also uses three winches. The cargo purchase is

276

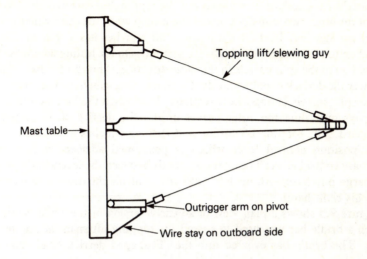

Topping lift/slewing guy

Mast table →

Outrigger arm on pivot

Wire stay on outboard side

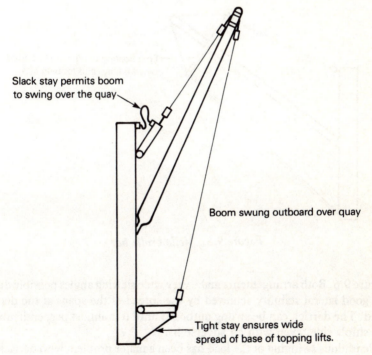

Slack stay permits boom to swing over the quay

Boom swung outboard over quay

Tight stay ensures wide spread of base of topping lifts.

Figure 9.4   Hallen outrigger rods

operated by a standard type winch but the topping lifts are arranged so that one of the other two winches controls the luffing while the third winch is used solely for slewing. Each of the topping lift winches has a split or divided barrel on to which the ends of falls are secured. On the luffing winch the falls are laid on to the split barrels in the same direction. Thus both falls will hoist or lower the derrick simultaneously. On the slewing winch the falls are laid on to the split barrels in opposite directions. Thus when the barrels rotate, one fall pays out while the other heaves in and the derrick slews to port or starboard. The topping lift luffing and slewing winches are operated by a multi-position control lever which is positioned adjacent to the cargo purchase control lever. The operator stands between the levers and operates the cargo purchase with his left hand and controls the derrick movements with his right hand.

Figure 9.5 shows a plan view of an early version of the Velle derrick in which a bridle bar was used to spread the topping lift spans at the derrick head. The bridle bar evolved into the 'T'-shaped derrick head shown in

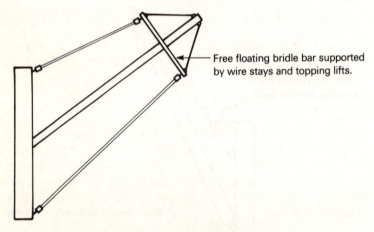

Free floating bridle bar supported by wire stays and topping lifts.

Figure 9.5   Velle bridle bar

Figure 9.6. Both arrangements make very wide slewing angles possible due to the good lateral stability achieved by the spread of the spans at the derrick head. The derrick can be swung outboard until it is almost perpendicular to the ship's side, even with an adverse list.

Pendulous swinging of the load has been a major problem with derricks in which the load hangs from a 'single point'. Good load stabilization is achieved with the T-shaped derrick head as the spread of the cargo runner reduces pendulous swinging and load rotation.

The Velle derrick is noted for its comparatively simple design, reliability,

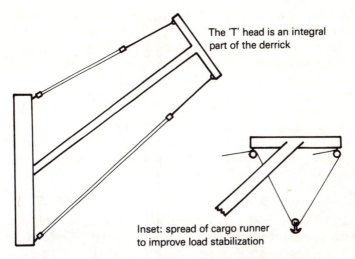

The 'T' head is an integral part of the derrick

Inset: spread of cargo runner to improve load stabilization

Figure 9.6   Velle 'T' derrick head

and versatility. The standard designs operate up to a capacity of approximately 35 tonnes but heavy duty designs are capable of lifting approximately 100 tonnes.

Stuelcken derrick

The Stuelcken derrick is probably the best known of all heavy lift designs. The 'fork-type' Stuelckenmast was first introduced in 1954 and since then many hundeds of Stuelckenmast derricks have been installed on many types of ship. There are five basic versions but Figure 9.7 shows a general arrangement plan of a 'typical' Stuelckenmast. All the Stuelcken derricks have common features:

1 The ability to swing through the vertical and 'plumb' two hatches without re-rigging, i.e. the working area is doubled as the one derrick can serve the hatches forward and aft of its position.
2 The twin V masts give a large distance between the base of the topping lift spans and thus a large horizontal arc of operation.
3 The twin span system means that slewing guys do not obstruct the cargo working area.
4 A portable remote-control device which is slung around an operator's neck means that the operator's view is never obstructed and gives a very large scope for automatic remote control.
5 There is no necessity to re-rig the derrick at any stage of the operation.
6 All blocks, swivels, goose necks, etc, are equipped with anti-friction

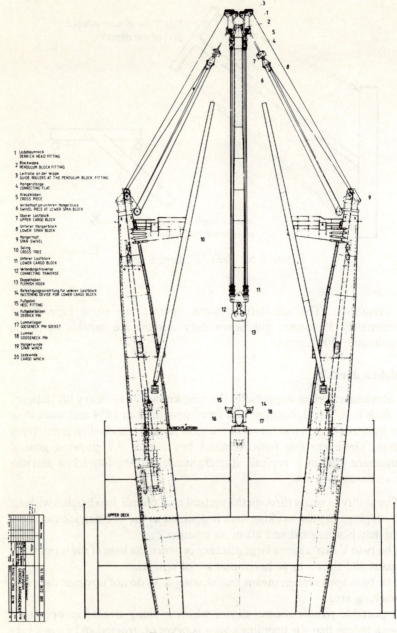

1 Ladebaumnock
  DERRICK HEAD FITTING
2 Blockwippe
  PENDULUM BLOCK FITTING
3 Leitrolle an der Wippe
  GUIDE ROLLERS AT THE PENDULUM BLOCK FITTING
4 Hängerstange
  CONNECTING FLAT
5 Kreuzkloben
  CROSS PIECE
6 Wirbeltopf am unteren Hängerblock
  SWIVEL PIECE AT LOWER SPAN BLOCK
7 Oberer Lastblock
  UPPER CARGO BLOCK
8 Unterer Hängerblock
  LOWER SPAN BLOCK
9 Hängertopf
  SPAN SWIVEL
10 Saling
   CROSS TREE
11 Unterer Lastblock
   LOWER CARGO BLOCK
12 Verbindungstraverse
   CONNECTING TRAVERSE
13 Doppelhaken
   FLEMISH HOOK
14 Befestigungsvorrichtung für unteren Lastblock
   FASTENING DEVISE FOR LOWER CARGO BLOCK
15 Fußgabel
   HEEL FITTING
16 Fußgabelbolzen
   DERRICK PIN
17 Lummellager
   GOOSENECK PIN SOCKET
18 Lummel
   GOOSENECK PIN
19 Hängerwinde
   SPAN WINCH
20 Ladewinde
   CARGO WINCH

Figure 9.7   Stuelckenmast

280

bearings and are sealed. Thus the derricks are practically maintenance-free for at least four years.

7 The derrick is secured for sea in a short period of time.
8 To reduce top weight there is much use of low-alloy high-strength steels.
9 There is very accurate spot loading capability.
10 Two derricks can combine their SWLs, e.g. two 130-tonne derricks can lift 260 tonnes.

Stuelckenmasts are built within the range of 20 tonnes SWL to approximately 500 tonnes SWL and the manufacturers have the capability of developing derricks of a greater capacity should the industry require such derricks.

Derrick cranes

Derrick cranes are difficult to define. Lloyd's Code for Lifting Appliances states that if a conventional derrick is fitted with one of the following modifications it is considered to be a derrick crane:

'(a) Twin span tackles so designed that the derrick can be slewed without the use of separate guys;
(b) A system for luffing the derrick boom other than by means of span ropes;
(c) The cargo and/or the span winches built into the derrick boom and moving with it;
(d) A system for slewing the derrick boom by applying a torque to a slew ring or trunnion'.

If more than one of the above modifications is fitted Lloyd's will normally consider the appliance to be a crane.

Technically, therefore, the derricks which have been previously discussed are crane derricks but most mariners still consider them to be derricks. In general, the derrick crane is suspended from a single topping lift which is controlled by its own winch. Slewing is achieved by means of a guy winch with a centre drum flange with two ropes running on to it from opposite directions. While one rope is hauled in and slews the crane derrick the other slacks out rope and stabilizes the appliance.

The main features of a crane derrick are:

1 The avoidance of high-level weight which is inevitable with a cross-tree or similar arrangement.
2 A level of cargo-handling efficiency comparable with that of a crane.
3 The use of standard and simple components reduces the capital and maintenance costs.
4 The absence of cross-trees and the use of simple components of

comparatively little weight lowers the position of the ship's centre of gravity which often gives greater stowage versatility.

Derrick safety devices

One of my methods of keeping 'up-to-date' is to visit ships in the port of Belfast and to discuss the operation of equipment with ships' officers. I have noticed that some Hallen and Velle type derricks are 'copies' and not patent and do not have the same safety standards as patent derricks. Mariners should be aware of the following:

1 Hallen derricks have built-in limit switches (somewhat similar to limit switches on lifeboat davits which prevent the lifeboat from being 'rammed home') on top of, and on the sides of, the derrick boom to prevent excessive topping and slewing. The switches can be in the form of a collar around the lower portion of the boom and if the collar touches anything the winches cut out and lock and thus require the operator to re-position the derrick.

2 Velle derricks have various safety devices including:
   (a) Slack cargo wire cut-out, e.g. if the lower cargo block rests on an object such as a wharf shed the winches lock
   (b) A limit switch (often suspended from the T bar) which prevents the lower cargo block from jamming hard against an upper cargo block by causing the winches to lock
   (c) Limit switches which prevent the derrick from over-topping, over-slewing or being operated at an angle too close to the horizontal; once again, the winches lock.

3 Some ships with non-patent Velle type derricks have split drum separator plates which are not deep enough and there is a tendency for the wire to jump from one part of the split drum to the other. Velle derricks have a deep divide on the split drum to prevent this and also a limit switch to ensure that there is a minimum number of turns on the drum (sometimes six turns).

4 A few officers have stated that it is sometimes difficult to keep the correct tension on wires on split drum assemblies and it is essential that each 'part' of the wire should exactly match the other. Some split drum wires are of one continuous length and thus if one part of the wire breaks the other part of the wire also slackens; some split drum wires, however, are in two separate parts.

5 Electric-hydraulic winch systems usually have a by-pass or loop which allows the oil to flow then the winches cut out and lock, thus keeping the oil cool.

6 Modern winches have features such as:
   (a) Three 'gear' settings—Neutral

   Speed 1 (faster)

   Speed 2 (slower)

(b) A 'step-by-step' mechanism which prevents the joystick control lever being moved too quickly
(c) An emergency stop button

Officers sailing on ships with patent or patent type derricks should be thoroughly familiar with all the safety devices on the derricks.

Officers with an interest in history may be amused to learn that Elizabeth the First's hangman, named Derrick, was so efficient that ship's cargo-handling appliances were named derricks after him.

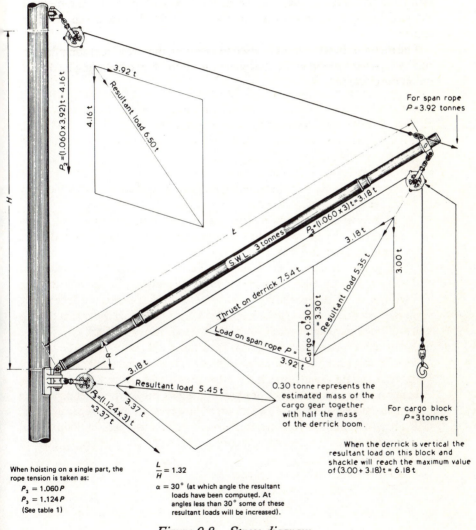

For span rope
P = 3.92 tonnes

For cargo block
P = 3 tonnes

0.30 tonne represents the estimated mass of the cargo gear together with half the mass of the derrick boom.

When the derrick is vertical the resultant load on this block and shackle will reach the maximum value of (3.00 + 3.18)t = 6.18 t

When hoisting on a single part, the rope tension is taken as:
$P_2 = 1.060\,P$
$P_3 = 1.124\,P$
(See table 1)

$\dfrac{L}{H} = 1.32$

$\alpha = 30°$ (at which angle the resultant loads have been computed. At angles less than 30° some of these resultant loads will be increased).

Figure 9.8   Stress diagram

283

Derrick stress

Instruction in methods of calculating stress in derricks is best carried out in colleges and not within the constraints of a book. All candidates for DoT Certificate of Proficiency and BTEC HND in Nautical Science should have a personal copy of the 'Code of practice for design and operation of ships' derrick rigs', BS MA 48: 1976, as the code shows the recommended procedures for calculating loads in derrick rigs. Figure 9.8 shows a typical illustration from the code. In order to prepare such load diagrams rope tensions are estimated by means of tables given in the code and, by using the method shown, similar estimations may be made with the derrick boom at other angles.

The method shown cannot be used to ascertain the stress in union purhase rigs. A separate section in the code gives the methods of calculating loads in a union purchase rig.

10

Certificates and Surveys

Merchant Shipping (Load Lines) Rules 1968. SI 1968 No. 1053

The rules came into operation on 21 July 1968 and contain the requirements for the surveying and assignment of freeboards to ships and the issue of load line certificates. The requirements for the marking of load lines are also specified and the Zones, Areas, and Seasonal Periods which are applicable to all ships to which the Rules apply are given in Schedule 2.

These rules enable the United Kingdom to give effect to the IMO 'International Convention on Load Lines 1966'.

Amendments to the 1968 rules are issued when necessary. SI 1979 No. 1267 came into operations on 1 January 1980 and SI 1980 No. 641 came into operation on 9 June 1980.

The Assigning Authorities are the Department of Transport or Classification Societies such as Lloyd's Register of Shipping.

Application for the assignment of freeboards to a ship and for the issue of a load line certificate is made to an Assigning Authority by or on behalf of the owners. The application must be accompanied by plans, drawings, and specifications which relate to the design and construction of the ship. The ship is then surveyed and the Surveyor must ascertain:

(a) that the construction of the ship shall be such that her structural strength will be sufficient for the freeboards to be assigned;

(b) that the stability in all probable loading conditions will be sufficient for the freeboards to be assigned; and

(c) that the construction of sills, hatch coamings, closing appliances, ventilators, air pipes, cargo ports, scuppers, inlets and discharges, side scuttles, freeing ports and arrangements, arrangements for the protection of the crew such as guard rails, all comply with the specifications laid out in the Rules.

The Surveyor provides the Assigning Authority with a report which gives the results of the survey and if that proves satisfactory freeboards are assigned to the ship. The owner is then given particulars of the freeboards assigned and the positions in which load lines, the deck line, and the load line mark are to be actually marked. The owner is also given two copies of the Surveyor's report and is issued with a 'Load Line Certificate'.

A load line certificate is valid for not more than 5 years after the date of completion of the survey. The Department of Transport may cancel a certificate if:

1 The ship does not comply with the conditions of assignment.
2 The structural strength of the ship is lowered to an unsafe standard.
3 The information on which the freeboards were assigned was incorrect.
4 A new certificate is issued.
5 The ship ceases to be registered in the United Kingdom.
6 The ship is not periodically inspected.

Periodical inspections are carried out by a Surveyor to ensure that:
(a) the fittings and appliances for the protection of openings, the guard rails, the freeing ports, and the means of access to the crew's quarters are in an effective condition; and
(b) no changes have taken place in the hull or superstructure. The inspection shall be carried out within 3 months before or after each anniversary of the date of completion of the survey which led to the issue of the certificate. The intervals between inspections shall not be less than 9 or more than 15 months. After satisfactory inspection the Surveyor endorses a record of the inspection on the load line certificate.

Deck line

This is a horizontal line 300 mm long and 25 mm wide which is marked amidships in the shell plating on each side of the ship to indicate the position of the freeboard deck. It should be painted a distinctive colour and it is the position from which freeboard is measured vertically downwards as the upper edge passes through the point where the upper surface of the freeboard deck intersects the shell.

Freeboard deck

This is the uppermost continuous deck below which water-tight integrity can be maintained, i.e. it is that deck on which all openings have permanent water-tight closing arrangements and below which all openings in the shell also have permanent water-tight closing arrangements. It is usually the weather deck, i.e. the deck exposed to the sea and weather, but in certain circumstances can be the deck below the weather deck.

Freeboard

This is the distance from the waterline to the upper edge of the deck line measured vertically, or with regard to the rules, 'the distance measured vertically downward amidships from the uper edge of the deck line to the position at which the upper edge of the load line appropriate to the freeboard is to be marked'.

Minimum freeboards are assigned so that a vessel remains seaworthy when loaded and to provide reserve buoyancy so that:

(a) the vessel will not be in danger of foundering in heavy seas;
(b) in the event of major damage the vessel will still remain afloat or will sink slowly enough so as to enable the crew to get clear.

Load line zones

The 'Zones, Areas and Seasonal Periods' chart which is located at the back of the 'Mercant Shipping Load Line Rules' should be studied carefully. Freeboards are correlated carefully with the weather conditions throughout the world and thus ships trading in areas which have poor weather conditions are loaded so as to provide greater freeboards. A larger freeboard provides more reserve buoyancy and the greater height of the deck above the sea lessens the impact of large seas on the ship.

There are two permanent zones, 'Tropical' and 'Summer', where weather conditions are similar throughout the year. 'Seasonal Areas' are areas within zones which have seasonal weather conditions which are somewhat different from the weather conditions which pertain to the whole zone. The 'North Pacific Seasonal Tropical Area' which is located on the northern boundary of the Tropical Zone is Tropical from 1 April to 31 October and Summer from 1 November to 31 March.

'Seasonal Zones' are zones in which weather conditions can vary greatly between seasons. In the 'North Pacific Winter Seasonal Zone' ships may load to a Summer freeboard between 16 April and 15 October but from 16 October to 15 April the minimum permitted freeboard is Winter. In certain zones the ship's length affects the permitted freeboard, i.e. the Baltic is always a 'Summer Zone' for ships over 100 m in length but it is a 'Winter Seasonal Area' for ships of 100 m or less in length. A vessel of 100 m or less in length loading within the Baltic from 1 April to 31 October may load to a Summer freeboard but from 1 November to 31 March the minimum freeboard is Winter.

One of the Chief Officer's functions is to load the maximum possible tonnage while ensuring before departure that at no time during any part of a voyage will the applicable seasonal mark be submerged. It is essential to remember that the vessel is governed not only by the zone in which she loads, but also by the zones through which she must sail. A vessel may load in a Tropical Zone but if the course takes the ship through a Summer Zone, the Summer mark must not be submerged while the vessel is in that zone.

The Chief Officer must therefore check which zone will be the 'governing zone' on any passage, i.e. the governing zone indicates the minimum permitted freeboard for that part of the passage. If the loading zone permits a lesser freeboard than a controlling zone, the Chief Officer first checks the permitted tonnage when entering the controlling zone and 'works backward' to the loading port by calculating the bunker and fresh-water consumption while sailing to the controlling zone. Such consumption is known as the 'Zone Allowance', which is the extra tonnage a vessel may load beyond that permitted by the controlling zone.

Preparations for a load line survey

Particulars relating to the conditions of assignment can be found in a form, commonly known as the 'Load Line Record', which is carried on board. The Chief Officer should study the record carefully to ensure that all the appropriate particulars have been incorporated into the 'Planned Maintenance Schedule'. The detailed preparation should commence three months before the expected date of the survey.

1 Check that all access openings at ends of enclosed structures are in good condition. All dogs, clamps, and hinges should be free and well greased. All gaskets and water-tight seals should be crack free. Ensure that the doors open from both sides.
2 Check all cargo hatches and access to holds for weathertightness, especially battening devices such as cleats and wedges.
3 Check the efficiency and securing of portable beams.
4 If portable wooden hatch covers are used, check that they are in good condition and that the steel binding bands are well secured.
5 If tarpaulins are used at least two should be provided for each hatch. The tarpaulins must be in good condition, waterproof, of ample strength, and of an approved material.
6 Hatches which are closed by portable covers and made weather-tight by tarpaulins must have a steel locking bar across each section. Covers more than 1.5 m in length must be secured with two locking bars.
7 Inspect all machinery space openings on exposed decks.
8 Check that any manholes and flush scuttles are capable of being made water-tight.
9 Check that all ventilator openings are provided with efficient weather-tight closing appliances and repair any defects.
10 All airpipes must be provided with permanently attached satisfactory means for closing openings.
11 Inspect any cargo ports below the freeboard deck and ensure that all of them are water-tight.

12 Ensure that the non-return valves on overboard discharges are operating in a satisfactory manner.

13 Side scuttles below the freeboard deck or to spaces within enclosed superstructure must have efficient internal water-tight deadlights. Inspect the deadlight 'rubbers'.

14 Check that all freeing ports are in a satisfactory condition, e.g. shutters are not jammed, hinges are free, and that pins are of non-corroding material. Check that any securing appliances, if fitted, work correctly.

15 All guard rails and bulwarks should be in a satisfactory condition, e.g. all fractured rails should be re-welded.

16 If life lines are required to be fitted in certain areas, rig the lines and overhaul as necessary.

17 De-rust and paint the deck line, load line mark, load lines, and the draught marks.

In brief, ensure that the hull is water-tight below the freeboard deck and weather-tight above the freeboard deck.

On the day of the survey have the Certificate and Record ready for the Surveyor's inspection. The Master should have sufficient stability information to show that the vessel can be loaded and ballasted correctly. Have all the necessary keys for areas which the Surveyor may want to inspect, e.g. store rooms. Sufficient men should be available for work such as opening cargo hatches, and ladders and stages should be ready for the Surveyor to view the load line marks.

Merchant Shipping (Cargo Ship Safety Equipment Survey) Regulations 1981. SI 1981 No. 573

These regulations came into operation on 1 May 1981 and apply to all United Kingdom ships (except passenger ships, fishing vessels, and pleasure craft) of 500 gross tons or over engaged in international voyages.

An application for a survey must be made to the Secretary of State by the owner and the application must be accompanied by sufficient information to enable the survey to be conducted properly. The survey is usually carried out by a Department of Transport Surveyor who must satisfy himself that the ship complies with the relevant Safety Regulations and that the equipment is satisfactory for the service for which the ship is intended. The Surveyor sends the Secretary of State a 'Declaration of Survey' which contains particulars of the ship and her equipment and which enables a certificate to be issued.

The owner of every ship to which a certificate has been issued must cause a survey to be carried out within three months before or after the anniversary date of the certificate. The survey is termed an intermediate survey in respect of tankers of ten years of age and over and an annual survey in respect of other ships.

M963 specifies the procedures for the survey and officers should be conversant with the notice.

Preparations for a cargo ship safety equipment survey

All safety equipment should be kept in excellent condition and a 'Planned Maintenance Schedule' which includes the safety equipment will ensure that a satisfactory survey can be carried out at any time. However, it may be desirable to overhaul the equipment prior to an expected survey. M963 states that part of the annual survey should consist of 'a visual examination of sufficient extent, together with certain tests of the ship's safety equipment, to confirm that its condition is being properly maintained'. The ship's certificates will also be examined and equipment will be checked to ensure that no unauthorized modifications have been made. The stringency of the survey will depend upon the condition of the ship's safety equipment. The Official Log Book will also be examined to establish whether the 1986 Muster Regulations have been adhered to.

The following preparations should be carried out shortly before the expected survey date:

1 Inspect all the lifeboat stores and equipment. Overhaul and renew as necessary.
2 Inspect the lifeboats, pay particular attention to buoyancy material and check that bottom boards and thwarts are not cracked. Repaint the ship's name, port of registry, and the lifeboat numbers, and ensure that the lifeboat particulars on the bow have not been obliterated.
3 Thoroughly overhaul davits, winches and blocks, and grease all moving parts. Re-new or 'end for end' the falls. Inspect lifeboat embarkation arrangements and launching arrangements and lower the boats into the water.
4 When the boats are in the water run any lifeboat engines both ahead and astern.
5 Check that the inflatable liferafts have been serviced within the previous 12 months. Inspect the stowage, release, launching and embarkation arrangements of the liferafts and, if necessary, re-new the launching instructions. Inspect any rigid liferafts.
6 Inspect the survival craft portable radio equipment.
7 Overhaul the lifebuoys, especially the self-igniting lights and self-activating smoke signals, and ensure that the lifebuoys are correctly located throughout the ship.
8 Examine the lifejackets and check that they are correctly distributed throughout the vessel.

9 Ensure that all pyrotechnics, including the line-throwing appliance rockets, are not out of date. Inspect the line-throwing appliance.

10 Test the emergency lighting at one of the times that the general alarm system is tested.

11 Check that the fire control plans are still posted and clearly legible.

12 Test, where possible, the fire and/or smoke detection system.

13 Run each fire pump, including the emergency fire pump, to check that each pump can supply via the fire main the required two jets of water simultaneously from separate hydrants.

14 Check that fire hoses, nozzles and applicators are in good condition and correctly located.

15 Test and overhaul the fixed fire-fighting system. Note that the instructions are posted and that controls and pipes are correctly marked.

16 Overhaul portable and non-portable extinguishers and check the securing arrangements. Ensure that extinguishers are correctly located and that spare charges are available.

17 Where possible, confirm that all remote controls are operable.

18 Overhaul any applicable closing arrangements for ventilators, skylights, doorways, funnel spaces, and the tunnel.

19 Overhaul the fireman's outfits and re-charge, when possible, the compressed air cylinders.

20 Inspect the pilot ladders, pilot hoist if carried, and all ancillary equipment.

21 Remember that the navigation equipment is also surveyed, e.g. navigation lights, shapes, sound signalling equipment, daylight signalling lamp, radar, echo-sounder, gyro-compass, and direction finder. The compass deviation book, charts and the publications that are required by the Merchant Shipping (Carriage of Nautical Publications) Rules 1975, SI 1975 No. 700, are also inspected.

Tankers have an additional survey which covers:

1 The piping of the fixed fire-fighting system of the cargo pumproom.
2 The deck foam system and the deck sprinkler system.
3 The inert gas system.

The intermediate survey of a tanker of ten years and over should include as a minimum all the relevant items mentioned above but it should also be 'sufficiently extensive to ensure that the ship's degree of compliance with the cargo ship safety equipment certificate warrants the ship's continued possession of that certificate and that the ship can continue to be operated with safety' (M963, Annex II).

On the day of the survey the relevant certificates and publications should be gathered together in one location for ease of examination. Lifejackets should not be assembled in one place but should be left distributed throughout the ship in the normal stowage positions. Ensure that lifejacket donning instructons have been posted. Check that the international shore connection, nozzles, etc., have not been appropriated by shore personnel. A sufficient number of crew members should be available to assist the Surveyor.

Merchant Shiping (Cargo Ship Construction and Survey) Regulations 1981. SI 1981 No. 572.

These regulations came into operation on 1 May 1981 and in general apply to seagoing United Kingdom ships of 500 gross tons or over (except passenger ships, troopships, pleasure craft, fishing vessels, and ships not propelled by mechanical means).

The owner of every ship to which the regulations apply must arrange for the ship to be surveyed on completion and thereafter at intervals not exceeding 5 years. Application must be made to a Certifying Authority, such as Lloyd's Register of Shipping, who arrange for the ship to be surveyed.

The Surveyor must satisfy himself that the arrangements, materials and scantlings of the structure, boilers and other pressure vessels, main and auxiliary machinery (including steering gear), electrical installations, and other equipment comply with the regulations. In the case of tankers the outside of the ship's bottom, the pump rooms, cargo and bunker piping systems, vent piping, pressure vacuum valves, and flame screens must also be inspected. The Surveyor sends the Assigning Authority a 'Declaration of Survey' which contains the necessary particulars of the ship and which enables a certificate to be issued.

The owner of every ship to which these regulations apply must arrange for the ship to be periodically surveyed in the manner specified in regulation 72(3) and (4) with regard to ship side fittings (such as overboard discharge valves), boilers, screw propeller shafts, and tube shafts.

The owner of every ship to which Part III of the regulations apply, i.e. tankers, must arrange for an additional survey not more than 6 months before, nor later than 6 months after, the half-way date of the period of validity of the certificate. This intermediate survey deals with the material and equipment in, or associated with, fire-retarding bulkheads.

In addition, the owner of every tanker of ten years of age and over shall arrange for an intermediate survey at least once during the period of validity of the certificate. If only one such intermediate survey is made, it must be carried out not more than 6 months before, nor later than 6 months after, the half-way date of the period of validity. This intermediate survey must be in accordance with the procedures specified in M1134 which states that in addition to the requirements for the annual survey noted below, an examination of shell plating, sea connections and overboard discharges, anchors and mooring equipment, the interior of at least two cargo tanks, rudder bearing clearances, propeller and shaft, boilers, electrical equipment in hazardous zones, and deck piping must be carried out.

Regulation 74 states that all ships which have a cargo ship safety construction certificate shall be subjected to an annual survey which must be carried out within 3 months before or after the anniversary of the certificate. The survey should be carried out in accordance with the procedures specified in M1134. In brief, the Surveyor will examine:

1 The ship's certificates.
2 The hull and closing appliances.
3 Anchoring and mooring equipment.
4 The operation of water-tight doors.
5 Water-tight bulkheads
6 Structural fire protection arrangements.
7 The operation of fire doors.
8 The machinery and electrical installations such as the propulsion system, steering arrangements, bilge pumping systems, boilers, and emergency sources of power.

In addition to machinery records, the Official Log Book will be examined to establish that the steering gear has been tested in accordance with the Merchant Shipping (Steering Gear and Automatic Pilot Testing Procedures) Regulations 1981.

Tankers will also have a weather deck survey in which the cargo tank openings, pressure-vacuum valves, flame arresting screens, piping, and electrical appliances will be examined. The cargo pump rooms will also be surveyed.

The above construction regulations incorporate the tanker steering gear arrangements promulgated by the 78 SOLAS Protocol.

The Merchant Shipping (Cargo Ship Construction and Survey) Regulations, SI 1984 No. 1217, apply to ships built after 1 September 1984. The survey of such ships is the same as that for older ships; M1134 gives the survey details.

Merchant Shipping (Steering Gear and Automatic Pilot Testing Procedures) Regulations 1981. SI 1981 No. 571

These regulations came into operation on 1 May 1981 and apply to seagoing United Kingdom ships. The regulations are the British equivalent of the new steering gear requirements of the IMO SOLAS Protocol of 78, the test procedures of which were referred to in Chapter 2.

Shipboard Operations

Officers should ensure that the tests and checks of the steering gear should include:

1 The full rudder movement.
2 A visual inspection of the steering gear and its connecting linkage.
3 The communication between the navigating bridge and the steering gear compartment.

Simple operating instructions, which include a block diagram showing the changeover procedures, for the steering gear control systems and power units must be permanently displayed on the navigating bridge and in the steering gear compartment.

The details of routine checks and emergency steering gear drills must be recorded in the Official Log Book. M1040 should be closely studied, in particular with reference to the use of an automatic pilot in areas of high traffic density.

Classification

Classification societies publish rules and regulations which set standards for ship construction and maintenance. When a ship is classified it is shown to be of sound construction and 'fit to do the job'. The oldest, and best known, of the classification societies is Lloyd's Register of Shipping. Lloyd's publishes *Rules and Regulations for the Classification of Ships* and all ships which are to be classed by Lloyd's must be built to the specifications contained in the rules. Constructional plans, the materials used, the methods of construction, and the standards of construction must all meet strict specifications. Lloyd's Surveyors attend the building of a vessel and conduct surveys throughout the life of the vessel to ensure that high standards are maintained. Lloyd's Rules state that 'ships built in accordance with the Society's Rules and Regulations will be assigned a class in the Register Book and will be continued to be classed as long as they are found, upon examination at the prescribed surveys, to be maintained in accordance with the Requirements of the Rules'. Most officers do not realize that they play a vital role in the maintenance of class as, 'The Rules are framed on the understanding that ships will be properly loaded and handled'.

Ships when classed are assigned one or more character symbols to denote their class, the highest class being ✠100 A1. These characters mean:

✠ New ships built under the supervision of the Society's Surveyors.
100 Considered suitable for seagoing service.
A Built or accepted into class in accordance with Lloyd's Rules and maintained in good and efficient condition.

1   Good and efficient anchoring and mooring equipment.

Any damage, defect, or breakdown, which could invalidate the classification conditions must be reported immediately to Lloyd's and all repairs which may be necesary for a vessel to retain her class must be carried out under the supervision of a Surveyor.

An Annual Survey must be carried out on all ships within 3 months, before or after each anniversary date of the completion of building, commissioning, or Special Survey.

Docking Surveys are carried out at intervals not exceeding 2 years, except that where high resistance paint has been applied to the hull the intervals may be extended to 2.5 years.

At Annual and Docking Surveys, the Surveyor examines the ship and machinery so that he may satisfy himself with regard to their general condition.

Special Surveys are carried out at 4-yearly intervals in a drydock, the surveys becoming more stringent as the age of the ship increases.

On the request of the owner a Continuous Survey may be carried out on the hull in which all compartments of the hull are opened for survey and testing in rotation, with a 5-year interval between examination of each part.

Complete Surveys of machinery are carried out at 4-yearly intervals. However, a Continuous Survey of machinery in which the various items are opened for survey in rotation may be conducted. In general, one-fifth of the machinery is examined each year with a 5-year interval between examinations of each item.

Senior Officers should read Part 1, Chapter 2, 'Classifications Regulations', of Lloyd's Rules for the details of particular surveys.

Survey requirements of SOLAS Protocol of 1978

IMO has published guidelines relating to mandatory annual surveys and unscheduled inspections of all cargo ships, and intermediate surveys on tankers of ten years of age and over, which are required by the Protocol of 1978. The United Kingdom has adopted the mandatory annual surveys in substitution for unscheduled inspections and the survey requirements of the Statutory Instruments previously discussed in this chapter meet the IMO requirements.

Code of Practice for Noise Levels in Ships

The Department of Transport issued this code in 1978 and M1305 draws attention to the main points of the code. The aims of the code are to limit maximum noise levels, to reduce exposure to noise, and to reduce noise levels generally.

The code applies to new ships over 24.4 m (80 feet) in length and to new ships below 24.4 m in length where reasonable and practicable. Certain sections which relate to potentially hazardous noise levels apply to existing ships.

When new ships are ordered due regard should be given to the code. Section 9 gives some methods for controlling noise exposure.

On new ships, i.e. contracts for which were signed after the publication of the code, a full noise level survey should be carried out as soon as practicable. On existing ships noise level measurements should have been taken as soon as practicable. The code contains information on requirements for noise surveys and survey reports, the latter being forwarded to the Department of Transport. The code also contains information on ear protection, training, and the responsibility of shipowners and seafarers.

11

Examination Preparations and Techniques

Many mariners fail examinations because they cannot lay out their answers in a manner which conveys to the examiner their knowledge and expertise in a particular subject. Learning 'by rote' is not a desirable quality in a ship's officer. He must acquire and develop the analytical capability which is required in modern industry. The mariner must not view the learning process as an academic exercise necessary to obtain certificates which make promotion, and an increased income, possible. The learning skills acquired at college are essential if one is to become a competent officer. In college one is presented with information which must first be assimilated and comprehended, and then related to industrial situations. The same process applies on board ship when officers have to deal with new technology and regulations, or when moving from one type of trade to another, e.g. from tankers to general cargo ships.

Throughout a course of study the mariner must learn the 'pros and cons' of the different facets of his courses and develop an evaluating approach to study. One must be able to discuss the advantages and disadvantages of a particular situation and to arrive at a reasoned conclusion. Students must develop the art of correct reasoning and show the examiners that they have such an ability. Memorizing rules and regulations is not enough; students must consider the reasons for the various regulations. The nature of our profession is such that we must train for the unexpected, both at sea and in the examination room.

Revision

Do not concentrate unduly on topics which you enjoy reading about. Revise carefully the topics you dislike as this is often a sign that you do not understand the essentials of the that topic. It is often useful to write brief notes or definitions on postcards which can be read on journeys or at odd moments during the day. Revise in a sensible manner; 'blockbusters' into the 'wee small hours' seldom give the desired results as tired minds do not assimilate information. Break off at the end of each hour for five minutes to give the brain time to rest. Do not read books in a mechanical manner but actually think about and evaluate the information that you are reading.

Before re-reading your topic notes, write down a brief summary of the subject and ensure that you fully grasp the most pertinent points. When you have read notes, ask yourself questions and write out the answers.

Examination preparations

Many mariners fail examinations because of an inability to understand the questions and not through lack of knowledge. You must be absolutely sure that you are answering the questions that have been asked by the examiner.

There is no substitute for attempting test papers under examination conditions and this should be regarded as an essential element of examination preparation. It is impossible to ascertain how well you can answer questions unless you attempt them without external help, i.e. by not referring to notes or to fellow students. No one enjoys attempting test papers under examination conditions but you cannot properly evaluate your revision requirements without doing so. The actual examinations will appear less formidable after a few 'mock' examinations. All colleges prepare test papers to examination standards and specifications and students should listen carefully to lecturers' criticisms of test papers. Most students who ignore the advice of lecturers fail the Department of Transport examinations.

Read the instructions at the head of the paper to see how many questions must be answered and note the overall time permitted for the examination. One aspect of an examination is to ascertain whether you have the ability to extract the central points from the knowledge you have acquired and to present the points in a manner which is relevant to the questions. Therefore, each answer should be well planned and set out in a logical manner. When possible start the answer with a clear statement that shows you understand the question, present your 'case' in the body of the answer, and finish with a strong conclusion. If a question is set out in sub-sections, e.g. (a), (b), and (c), lay out your answer in the same way. Do not answer (c) first as it shows an untidy and illogical approach which may confuse yourself and annoy the examiner. Do not waste time by writing out the question; the examiner knows what the question is, he requires an answer to it.

Rely heavily on common sense, be neither too brief or too verbose. Keep to the point, avoid generalizations and do not waffle or pad out the answer. You will only be awarded marks for points which answer the question; a brilliant answer on a different topic will gain no marks at all. Do not attempt to 'bluff' the examiner or try to fool him into believing that you know more about the subject than your answer indicates. It will not work; the examiner is a very experienced mariner, both operationally and academically, and has 'seen it all before'. A four-line answer and the terse statement, 'I could write more but I have not time to finish' does not impress.

Before answering, decide exactly what the examiner wants. Most questions contain a 'key word' or phrase which indicates the nature of the required answer. In one question the examiner may simply require facts, in another question you may be required to analyse, examine, or evaluate and give a reasoned conclusion. Some examples of key words or phrases are given below.

Describe   A factual description of characteristics is required. Thus a description of a pilot ladder would include the characteristics (material, size, etc.) of steps, spreaders, and side ropes.

Explain   This often requires analysis, reason, and clarification. If one had to 'explain' how a pilot hoist works the emphasis should be on the operational aspects and motive power of the hoist and not on a factual résumé of the pertinent regulations.

Describe and explain   A combination of fact and reason. 'Describe a pilot ladder and explain the uses of the component parts' would require a student to state the size of a spreader with the explanation that it prevents the ladder twisting.

Compare   This requires students to expound on the similarities of two subjects. Thus if one is required to compare epoxy and polyurethane paint a basic similarity would be that they are both two-pack paints.

Contrast   Emphasize the differences between the subjects.

Compare and contrast   To obtain good marks the student must point out the similarities and differences between the subjects. 'Compare and contrast epoxy and polyurethane paint' would require a student to mention that both paints produce a similar hard surface but that epoxy usually has a matt finish while polyurethane has a high gloss finish.

Discuss   Similar to 'explain' but usually there are several reasons for and against both sides of an argument. If one has to 'discuss the advantages and disadvantages of flame cleaning and grit blasting' one should examine, analyse and weigh up the various aspects of plate preparation and give a reasoned opinion as to which is the best method.

Define A concise statement. Thus 'define angle of repose' requires an authoritative meaning and nothing else.

Account for This usually deals with causes. An initially excessive angle of repose would be one of the factors to consider if one has to 'account for a list on a vessel carrying a bulk cargo'.

Enumerate or list Simply write concisely, in outline form the pertinent points relating to the question. 'List the factors one would consider when inspecting the forepeak' requires a mention of enclosed space procedures but not a page on the rescue procedures for removing an unconscious man from the forepeak.

Short notes This can often be a four-part question and each part of the answer should be approximately the same length, the total answer being of essay length. Thus each section should only be a quarter of the length of an 'average' answer and students should be careful to avoid writing an essay on one part of the answer.

In the 'Shipboard Operations' paper all answers should be written from the point of view of a Chief Officer. If a question requires you to give the Bosun instructions, do so in an authoritative way; do not request the Bosun in the manner of a first trip Third Officer. If a question asks, 'What action would you take?' imagine yourself to be a Chief Officer and proceed in the manner which you would expect a good Chief Officer to use.

When you have completed an answer re-read the question and ensure that you have not left any portion unanswered. Every word in a question should be carefully examined as it is there for a purpose.

On board ship one should behave in a 'seamanlike' way, use the same approach in examination preparations and you will not go far wrong.

Examination techniques

It is essential that all the questions that require answering are attempted. It is very difficult to achieve full marks in essay answers; thus you will receive more marks in total by answering all the required questions. If the total marks in a paper are 100 and the number of required answers is 5, equal value questions will therefore be worth 20 marks each. If the required standard is 70 percent it will be extremely difficult to pass the paper if one question is omitted. The student will then be required to obtain 70 marks out of a possible 80, instead of 70 out of a possible 100 marks.

Divide up the examination time before commencing to write so that you know how much time to allocate to each question. Allow time to read over the paper and for reading through the answers. This should be done before leaving the examination room. A typical examination paper could consist of 6 equal value questions with the paper being required to be completed in $2\frac{1}{2}$ hours. If you allow 10 minutes at the beginning for reading the paper and organizing your plan of attack, and 10 minutes for reading the answers, you are left with approximately 20 minutes for answering each question. Thus the answers must be concise and only deal with the pertinent points; there is no time for vagueness or generalizations.

The following course of action should be adopted. It is a tried and tested method which can be used in most examinations. Mariners are professionals; do not expect to obtain professional qualifications by amateur methods.

1 Read all the questions and make sure that you thoroughly understand what the examiner is expecting from you.
2 Make out a plan and keep to it. Before the examination started you should have requested a sheet of paper from the invigilator. Write at the top of the sheet 'Rough notes—not to be marked' and hand the sheet in with your answers. Write the question numbers on the sheet and briefly note the essential points which you must refer to when answering each question. This is time well spent. It is difficult to move from one subject to another and the notes will help you to start the next question. If you have been discussing 'the problems of transporting concentrates in bulk' and are moving on to 'describe the technique of crude oil washing' a note of the pertinent features of the COW system will enable you to 'switch gear' quickly. If you have been foolish enough to have left insufficient time to deal with the last question adequately, the notes will help you to make some valid points in the last few precious minutes of the examination.
3 Start with a question which you know you can answer well. It is often difficult to get actually started and by answering a question with which you can cope you build up confidence for answering the rest of the paper.
4 Leave a question to the end which you know you can answer well. This requires a lot of willpower but you are more likely to keep to a schedule and to keep sufficient time for the final answer if you know that you can do well in the question.
5 Start each question on a new page. This enables you to add any important point which you may have forgotten when answering a previous question.
6 When writing one question, if you think of an important point which refers to another question make a note of it on your 'Rough Notes' page. Otherwise you may not remember until after you have left the examination room.
7 Read each answer as soon as you have completed it to ensure that you have given the desired information to the examiner.

Attempt to develop your answers logically and write in a direct simple manner. Avoid rhetoric and literary embellishment. Write legibly; the examiner cannot give you marks if he cannot read your answer.

Do not attempt to 'cram' the night before an examination. After months of preparation a few hours of panic studying will not help. Go to the cinema and relax.

Remember, there are no 'easy' papers. An examination only appears to be easy when you have completed months of careful studying and preparation.

Index